INSTINCTS

by design

Instinctive Behaviors of Humans & Animals in Creation

E. Howard Callahan

First printing: September 2023

Master Books, P.O. Box 726, Green Forest, AR 72638
Master Books® is a division of the New Leaf Publishing Group, LLC.

ISBN: 978-1-68344-285-1

ISBN: 978-1-61458-799-6 (digital)

Library of Congress Control Number: 2023943544

Cover and interior design by Diana Bogardus

Please consider requesting that a copy of this volume be purchased by your local library system.

Printed in the United States of America

Please visit our website for other great titles: www.masterbooks.com

For information regarding promotional opportunities,
please contact the publicity department at pr@nlpg.com.

About the Author: E. Howard Callahan is a retired chemical engineer with a master's degree in theology. As a chemical engineer, he spent years working in the fields of chemistry and biology. As a Christian, he has spent over 60 years studying the Bible and how it applies so well to our everyday lives. He continued learning physics and chemistry, including organic and inorganic, as he got his bachelor's degree in chemical engineering at Georgia Tech. Howard's engineering experience has included research and development (R&D) work that revealed to him how God made the world. Howard ran the chemistry lab for Dr. O.A. Battista, who was an expert in polymer chemistry, especially in the two polymers that make up probably 80 percent of all living things: cellulose and collagen. Then Howard did R&D work on aloe vera, learning of its many uses in humans and extracting a drug from it. Next was R&D for an anticancer drug-delivery system. Lastly, Howard helped develop and manufacture an artificial scab for very large wounds.

INSTINCTS are a fascinating subject! We instinctively breathe and don't want to fall. When we look at the world, we see that many animals have instincts too.

What are instincts? How has God designed these features and passed them along to subsequent generations? What are some fascinating instincts that have been studied by researchers across the world?

This book sets out to be a detailed resource that looks at instincts like no other before it. But the uniqueness of this book is that it stands on God and His Word as the authority by which we look at the nature of instincts.

It's God's World—albeit suffering under the curse from sin in Genesis 3—but we can still see amazing remnants of God's original creation and aspects that built instinctive features animals now have. Join Howard Callahan as he takes you on a journey of discovery into the realm of instincts and the Bible.

Bodie Hodge
Speaker, writer, and researcher for Answers in Genesis

"Even the stork in the heavens
Knows her appointed times;
And the turtledove, the swift,
and the swallow

Observe the time of their coming.

But My people do not know the
judgment of the LORD."

Jeremiah 8:7

TABLE OF CONTENTS

The Elephant in the Room: The Unconscious Mind

People discussing a problem will occasionally avoid, on purpose or accidentally, an important part of the problem. When the ignored factor is really obvious and important for understanding or solving the problem, it is said that they are ignoring "the elephant in the room." For the problem of how did we get here, instincts are the elephant in the room.

God has made a proof of His being the Designer of life so obvious that humans often don't see it anymore. We see instincts at work all the time. We understand that we have little to no control over them. So what do we do? We say that they are controlled by our "unconscious mind" — and stop thinking about them.

Go to the ant, you sluggard!
Consider her ways and be wise,
Which, having no captain,
Overseer or ruler,
Provides her supplies in the summer,
And gathers her food in the harvest.

Proverbs 6:6–8

In the Bible, Paul says, *"Since the creation of the world His invisible attributes are clearly seen, being understood by the things that are made, even His eternal power and Godhead, so that they are without excuse, because, although they knew God, they did not glorify Him as God, nor were thankful, but became futile in their thoughts, and their foolish hearts were darkened. Professing to be wise, they became fools…" (Romans 1:20–22).*

Where is this "unconscious mind" that controls so much of our lives and even keeps us alive? God has placed these "instructions" in several places in our bodies. Most are in the bottom half of our brain. The top half, the cerebrum, could be called our "conscious mind." Below it is the cerebellum and the brain stem, which biologists divide into several pieces, such as the medulla and pons. The "conscious mind" has little control over them, but they are "essential to survival." Just the medulla regulates many bodily activities and produces important reflexes.[1] But "instructions" are also in other places, such as the vagus nerve and even in the cerebrum. This book will look at the amazing "unconscious mind" and the "simple" instincts it controls. As it does, thank God and honor Him for His astounding abilities.

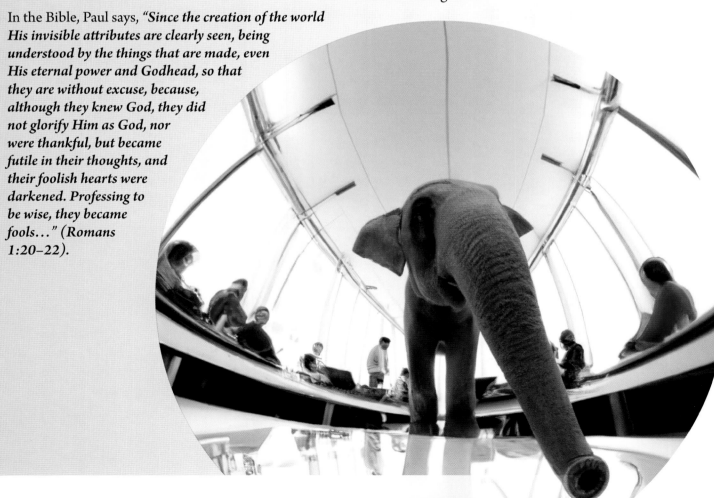

What Is an Instinct?

Instinct is a word used to describe a behavior that is built into the human, animal, or plant. One dictionary says an instinct is "an inborn pattern of behavior that is characteristic of a species…"[2] Almost always, it is a behavior necessary for the health, often the survival, of the creature. As you will see from this book, without instincts no life would exist.

Instincts are a major problem for evolution. The main reason is that each instinct is unquestionably a very complex set of instructions for a specific action in a specific creature, and it is hard to even imagine how any of these instructions could be made without a designer. Note this summary by an evolutionist: "Everyone in the world holds an inherent set of instinctive behaviors. Some of the behaviors include a collection of reflexes. Scientists continue to examine the purpose of these instincts and develop hypotheses on their function. **Many of the instincts remain an evolutionary mystery**"[3] (emphasis added). Most likely, the above emphasized phrase should be "**almost all** of the instincts remain an evolutionary mystery!"

The easiest way to see the problem for evolution is to imagine trying to program a robot to do all the steps necessary to duplicate an instinctive behavior. Each instinct is very complex, yet evolution claims that dumb luck (evolutionists will say "time, pressure, and chance, but no Intelligence") created every one of the thousands (possibly millions) of them! It is a ridiculous belief, yet the only way to explain it is if you say there is no God.

The goal of this book is not to "prove" there is a God, though creation itself does clearly testify about the Creator (Romans 1:20). Nor is it to say that evolutionists have not imagined plausible explanations for some of the almost innumerable instincts. The goal is to amaze you with the fact that almost nothing in life is really "simple," and the most logical, rational explanation is that God created everything.

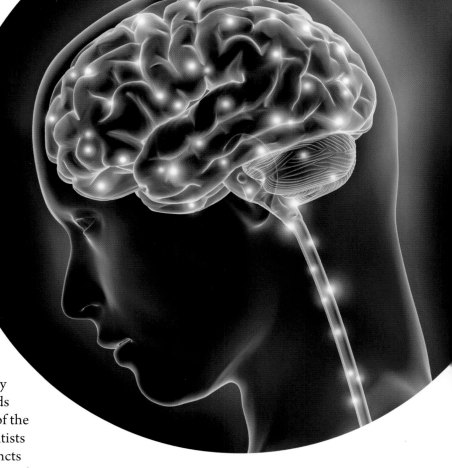

Albert Einstein, in a 1952 letter, described the universe in an interesting way:

> I am not an Atheist… We are in the position of a little child, entering a huge library whose walls are covered to the ceiling with books in many different tongues. The child knows that someone must have written those books. It does not know who or how. It does not understand the languages in which they are written. The child notes a definite plan in the arrangement of the books, a mysterious order, which it does not comprehend, but only dimly suspects. That, it seems to me, is the attitude of the human mind, even the greatest and most cultured, toward God. We see a universe marvelously arranged, obeying certain laws, but we understand the laws only dimly.[4]

This book will show you just a part of that huge library of laws, the instincts.

The Nobel Prize-winning Austrian zoologist Konrad Lorenz being followed by a group of ducklings. Lorenz studied medicine in Vienna before changing to zoology. His first discovery as a scientist occurred when he was given a one-day-old duckling that followed him around as if he were its mother. Together with the Dutch zoologist Niko Tinbergen, he founded a branch of animal behavior called ethology, based on observing the instinctive behavior of animals in the wild. They shared the 1973 Nobel Prize for physiology or medicine with Karl von Frisch.

What Do Evolutionists Say about Instincts?

Some evolutionists are presently implying that instincts are only learned. When you study what they have actually determined, you see that they found that some instincts (complex sets of instructions) are modified by experiences of the creature. The fact that some of the millions of instinctual actions are modified by "learning" gives the evolutionist hope he can believe there is no God. So, these few examples are presented to students with words that suggest further study will show less and less of the programming was actually built-in at birth.

> *DNA* **plays a critical role in these processes,** but does not by itself create traits. Accordingly, **instincts are not preprogrammed, hardwired, or genetically determined;** rather, they emerge each generation through **a complex cascade** of physical and biological influences.[5] (emphasis added to words of Mark S. Blumberg)

Notice how Blumberg admits that DNA is "critical" but not alone in creating the instincts he studied. Then notice how he refers to the **final version** of these instincts, so that he can say these final versions are not "preprogrammed, hardwired, or genetically determined" even though the **foundation** of even these instincts he studied are "preprogrammed, hardwired, and genetically determined." He even, indirectly, admits that these instincts which are often vital for survival are correctly created by "a complex cascade of physical and biological influences" for almost every one of the creatures of that species!

As you read through this book, notice that most of the instincts discussed could not have been modified by "a complex cascade" of events in the environment.

Another scientist has a more accurate summary of where true science is regarding instincts: "Although no one today seriously doubts that behavior is influenced in some way by **genetic constitution,** a general understanding of the mechanisms by which genes exert their influence is still faraway"[6] (emphasis added). Note that "genetic constitution" means at least some of the programming is built-in at birth.

Another evolutionist admits they are **"genetically encoded. Innate behaviors exist because they are necessary in some way for survival** and have **evolved through the process of natural selection**"[7] (emphasis added). Notice two ideas typically held by evolutionists. One, these very necessary behaviors exist because they are necessary. As you listen to nature programs, they will often say that some animal needed something, so it got it! They never say how they got it. They imply the animal DECIDED it needed it. This is supposedly answered by the idea that it evolved through natural selection.

Think about it. The word "selection" means you are choosing something from a group of things. The only group possible would be mutations of genes. As you read this book, you'll find no simple set of instructions (instincts). Evolutionists must BELIEVE that these "necessary in some way for survival" instructions built up over many generations, being useless or close to useless for all of those generations. I don't have that much faith!

Here are some telling quotes from scientists about instincts with my emphasis added.

"It's pretty clear that physical traits like the color of our eyes are inherited, but behavior is more complicated." Shook says, "It's a **complex interaction between genetics and environment.**"[8]

"Every organism is born with different biological traits and tendencies in order **to help them survive.** These aren't learned or experienced behaviors, rather patterns of behavior that occur naturally and are goal-directed. These patterns of behavior are referred to as instincts, and the theory suggests that **instincts drive all behaviors.**"[9]

"Animals react to things without thinking about it and the reaction they had was **completely functional the first time** they performed it."[10]

"Innate behaviors generally involve **basic life functions,** so it's **important that they be performed correctly.**"[11]

"**Instinct is a powerful force** in the animal world. It dictates the behaviors **necessary for survival,** especially in species that don't get much guidance from their parents. These behaviors are **programmed into an animal at a genetic level.** An innate behavior is inheritable, passing from generation to generation through genes. It is also intrinsic, meaning that **even an animal raised in isolation will perform the behavior,** and stereotypic, meaning that it is **done the same way every time.** Innate behaviors are also inflexible and are not modified by experience. Finally, they are consummate, which means that the behavior is **fully developed from the animal's birth.**"[12]

Which Instincts Will Be Investigated

Since God did not put any labels on creatures that would specify what were "instincts," it is up to humans to specify what is or is not an instinct. Because humans are not infallible, this book will use a very simple definition of instinct: ANY COMPLEX SET OF INSTRUCTIONS IN MOST CREATURES OF A SPECIES THAT WAS NOT LEARNED AND IS NOT THE RESULT OF CONSCIOUS EFFORT.

Due to the huge number of instincts, this book will not attempt to list all the different kinds, much less all the individual instincts. Instead, it will present some that are necessary for survival or growth or protection in humans. Crucial instincts are also in animals, so a few of theirs will also be presented.

Because many of the instincts God made are valuable for several purposes, it is difficult to categorize them. This book will group instincts, while recognizing some of them could belong in several groups. In fact, as in modern computers there are sets of programming code that are used for more than one purpose, so also in living creatures these instincts are often used for more than one purpose. This has caused this book to almost randomly shift between the singular "instinct" and the plural "instincts" when describing some marvelous behavior — we don't really know how many instincts are involved in a behavior!

This book will generally present instincts that are
1   necessary for life,
2   helpful for growth,
3   necessary for protection.

Let's begin examining the wealth of amazing instincts all around and in us. You should come away with an awe of the creative, perfectionist God who loves us so much that He did all of this for us. In talking about the human body, the psalmist ("poet") says it nicely: ***"I am fearfully and wonderfully made" (Psalm 139:14).***

"Science cannot solve the ultimate mystery of nature. And that is because, in the last analysis, we ourselves are part of the mystery that we are trying to solve."

— Max Planck

SECTION 1:

HUMAN

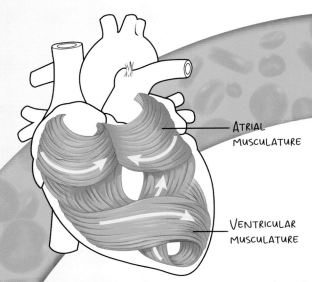

ATRIAL MUSCULATURE

VENTRICULAR MUSCULATURE

1

What is so important about your heart's pumping instinct?

Human Instincts Necessary for Life

God, in His wisdom, has made human babies incapable of many actions that they will later learn to perform easily, like walking and talking. However, God made babies capable of doing many things necessary for survival. Let's examine the things seen almost every day and probably not appreciated. In His great wisdom, God formed humankind, bestowing on us the gifts of instincts, empowering people with an innate wisdom and ability to survive and to thrive in this world.

Your heart is essentially a four-chambered muscle, working to pump blood all through the body. The Bible rightly says that *"the life of all flesh is its blood" (Leviticus 17:14).* Within a very short time, you would be dead if your heart stopped sending your blood throughout your body by pumping it. However, you would also be dead if your heart didn't beat at the right speed and with a good rhythm.

God has put a complex pumping instinct (set of instructions) inside the medulla. It is at work for you almost from the time of conception in the womb. It keeps working continually until almost the time of your death.

You have many muscles in your body, but almost none of them are moving continually or at a steady rate. Many of the muscles in your digestive system are actually pumps, but they move only when needed. God designed your heart to work, without stopping, for about 100 years! Nothing human scientists/engineers have ever made comes anywhere near this durability.

MEDULLA

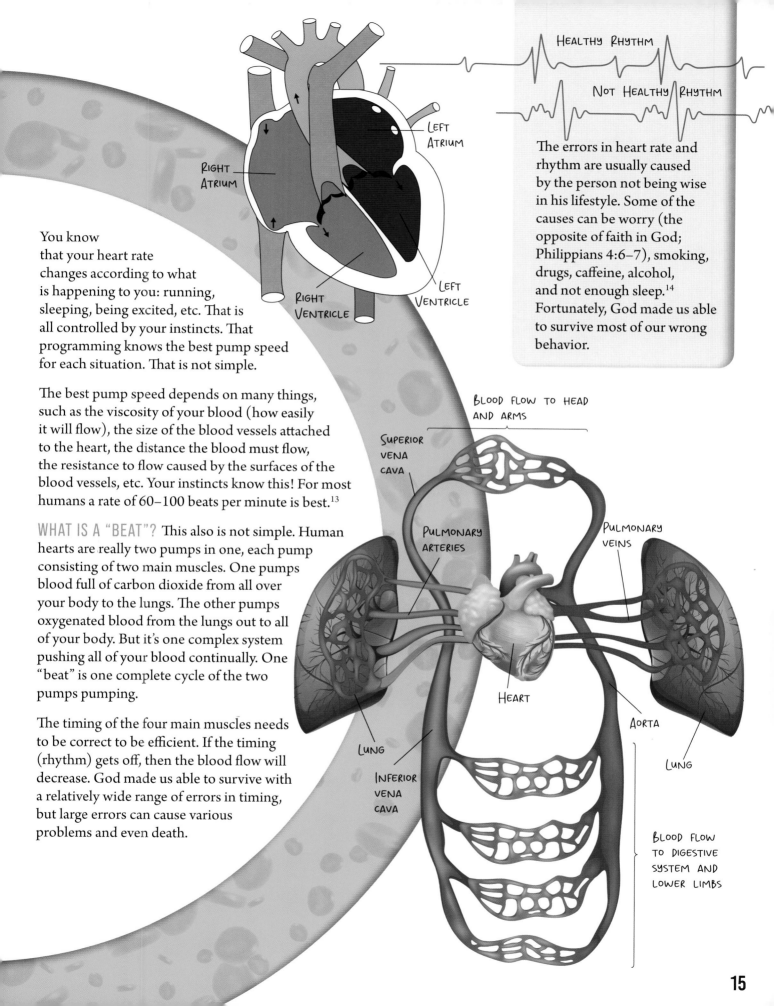

RIGHT ATRIUM

LEFT ATRIUM

RIGHT VENTRICLE

LEFT VENTRICLE

HEALTHY RHYTHM

NOT HEALTHY RHYTHM

The errors in heart rate and rhythm are usually caused by the person not being wise in his lifestyle. Some of the causes can be worry (the opposite of faith in God; Philippians 4:6–7), smoking, drugs, caffeine, alcohol, and not enough sleep.[14] Fortunately, God made us able to survive most of our wrong behavior.

You know that your heart rate changes according to what is happening to you: running, sleeping, being excited, etc. That is all controlled by your instincts. That programming knows the best pump speed for each situation. That is not simple.

The best pump speed depends on many things, such as the viscosity of your blood (how easily it will flow), the size of the blood vessels attached to the heart, the distance the blood must flow, the resistance to flow caused by the surfaces of the blood vessels, etc. Your instincts know this! For most humans a rate of 60–100 beats per minute is best.[13]

WHAT IS A "BEAT"? This also is not simple. Human hearts are really two pumps in one, each pump consisting of two main muscles. One pumps blood full of carbon dioxide from all over your body to the lungs. The other pumps oxygenated blood from the lungs out to all of your body. But it's one complex system pushing all of your blood continually. One "beat" is one complete cycle of the two pumps pumping.

The timing of the four main muscles needs to be correct to be efficient. If the timing (rhythm) gets off, then the blood flow will decrease. God made us able to survive with a relatively wide range of errors in timing, but large errors can cause various problems and even death.

BLOOD FLOW TO HEAD AND ARMS

SUPERIOR VENA CAVA

PULMONARY ARTERIES

PULMONARY VEINS

HEART

AORTA

LUNG

LUNG

INFERIOR VENA CAVA

BLOOD FLOW TO DIGESTIVE SYSTEM AND LOWER LIMBS

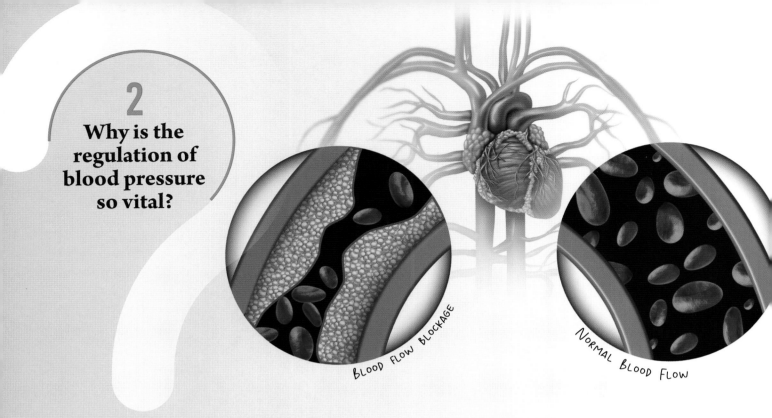

BLOOD FLOW BLOCKAGE

NORMAL BLOOD FLOW

2

Why is the regulation of blood pressure so vital?

One method that God's design uses to control the heart rate is blood pressure. When a pump like the heart is pumping, the liquid (blood) is pushed against the piping and other equipment downstream. When the piping can burst from too much pressure (like a blood vessel might), the pumping system needs sensors and instructions for what to do to prevent over-pressure (high blood pressure).

In a system like the human body, the blood circulates continually, so it needs to come back to the heart. If the returning blood has been slowed down for some reason, the heart (pump) will be forced to work harder to produce sufficient flow out of it. This happens in human bodies due to problems like hardening of the arteries and can cause many problems, including death.

How did God's design manage these potential problems? He built in a very complex set of instructions that will be described here in a much-simplified way.

YOUR HEART PUMPS ABOUT 1½ GALLONS OF BLOOD EVERY MINUTE. OVER THE COURSE OF A DAY, THAT ADDS UP TO OVER 2000 GALLONS.

Biologists know that the main source of the instructions (the "INSTINCT") for regulation of blood pressure and flow is the medulla oblongata in the brainstem.[15] There is a cluster of neurons in the medulla that respond to sensors spread throughout the body that monitor "changes in blood pressure as well as blood concentrations of oxygen, carbon dioxide, and other factors such as pH."[16] As you can see, the control is not simply blood pressure. The system is amazing.

? What do those blood pressure numbers 120/80 really mean?

Scientists have developed quick ways to check blood pressure and oxygen concentration, but even a baby's brain is doing that, and several other checks, to keep continuous control of his heart rate.

Doctors will measure our BLOOD PRESSURE with a device that detects the highest pressure in our arteries (called systolic; when the heart muscles are contracting, sending the blood out to the body) and the lowest pressure (called diastolic; when the heart muscles are expanding as they relax). The "normal" numbers are 120/80; they vary with what the body is doing, but they also vary with the health of the body. That is why doctors almost always want to see our blood pressure numbers.[17]

One of the sensors spread around the body that is used by the neurons in the medulla to control blood pressure and flow is called a baroreceptor.

"Baroreceptors are **specialized** stretch receptors located within thin areas of blood vessels and heart chambers that respond to the degree of stretch caused by the presence of blood."[18] Emphasis was added to this quote to show that a scientist saw the DESIGN without giving even the possibility that there was a Designer.

These baroreceptors are strategically located in your body. They are where a good Engineer would place them. They are where the blood leaves the heart to go to the body. They are where the blood is going to that most important part, the brain. They are in the vessels bringing the blood back from the upper body and from the lower body. They are also where the blood reenters the heart.

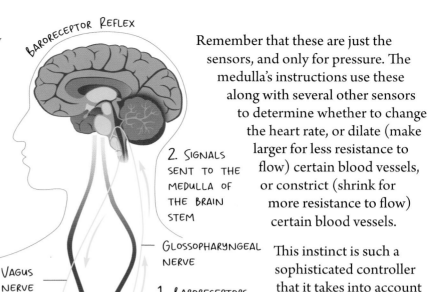

BARORECEPTOR REFLEX

2. SIGNALS SENT TO THE MEDULLA OF THE BRAIN STEM

GLOSSOPHARYNGEAL NERVE

VAGUS NERVE

1. BARORECEPTORS DETECT CHANGES IN ARTERIAL PRESSURE

3. HEART RATE ADJUSTS

Remember that these are just the sensors, and only for pressure. The medulla's instructions use these along with several other sensors to determine whether to change the heart rate, or dilate (make larger for less resistance to flow) certain blood vessels, or constrict (shrink for more resistance to flow) certain blood vessels.

This instinct is such a sophisticated controller that it takes into account not only blood pressure, but many other factors and body mechanisms. As our conscious mind does good and bad things to our bodies, this instinct adjusts. It adjusts for things like "**diet, exercise, disease, drugs or alcohol, stress, and obesity.**" It even adjusts as the limbic system (regulation of thirst, hunger, mood, etc. by the thalamus and hypothalamus in the brain) reacts physically to "psychological stimuli, chemoreceptor reflexes, generalized sympathetic stimulation, and parasympathetic stimulation."[19] God makes this instinct's complicated system work even if your conscious mind has no idea it's helping you.

AN ADULT HEART IS ABOUT THE SIZE OF 2 HANDS CLASPED TOGETHER

A CHILD'S HEART IS ABOUT THE SIZE OF A FIST.

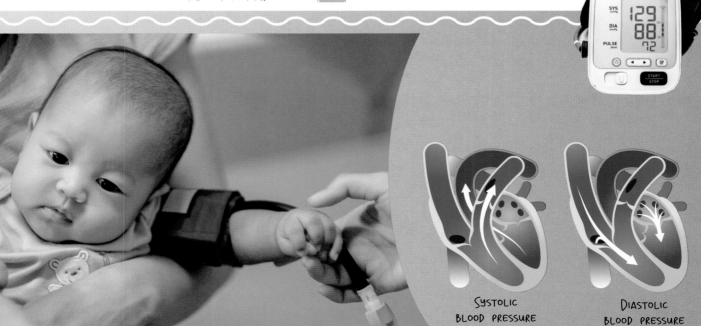

SYSTOLIC BLOOD PRESSURE

DIASTOLIC BLOOD PRESSURE

3

How does a baby know to inhale and exhale?

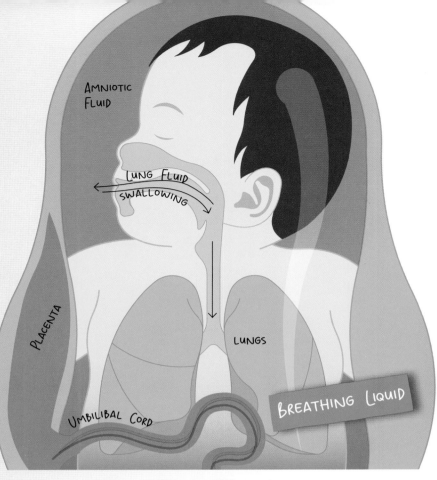

Breathing Liquid, Then Only Air:
Have you ever considered the fact that for 9 months prior to coming out and taking his first breath of air, a baby was living inside a fish bowl, submerged the whole time! While in the womb, the mother's blood provided the needed oxygen and removal of carbon dioxide. There was no need for the baby to breathe. Does the baby "breathe" inside the womb? No.

At about week 10 (of the 40-week time in the womb) the baby's lungs begin to develop by "inhaling" a little of the amniotic fluid he's swimming in. The first real inhaling (of the amniotic fluid) occurs at around week 32. It's apparently only "practice," since it doesn't do anything for the baby's body other than further develop the lungs' ability to perform "breathing." By week 36, the baby's lungs are fully developed.[20]

When biologists learn about living creatures, they often find that more than one important result comes from a single action. That is the case with a mother's contractions as her baby begins to come out of her

womb. The squeezing moves the baby into the correct position to come out. It also squeezes the baby's lungs to remove almost all of the amniotic fluid in them.

When the baby emerges, two important changes must happen.	
1	Within about a minute, the baby must begin inhaling and exhaling a gas (air), not a liquid, as it had been doing for the previous 8 weeks.
2	It must know to never inhale liquid again! This change has to occur within about one minute or the doctors consider it an emergency![21]

The baby cannot understand this, but his instincts take care of every step at the right times.	
Inside	**Outside**
Inside the womb, his body's oxygen/carbon dioxide swapping was indirectly handled by the mother's lungs breathing air. The umbilical cord allowed her blood to exchange them with his blood.	Outside the womb, the only oxygen/carbon dioxide swapping would have to be done by using his own lungs with air. Human lungs cannot perform the oxygen/carbon dioxide exchange with any liquid, only air.

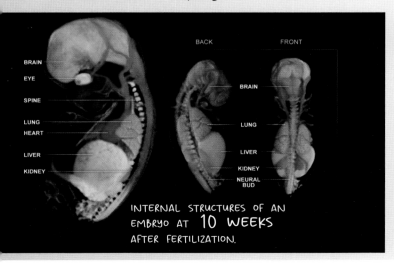

INTERNAL STRUCTURES OF AN EMBRYO AT **10 WEEKS** AFTER FERTILIZATION.

The instincts controlling all of this are amazing. Let's look at some details of the system of breathing air.

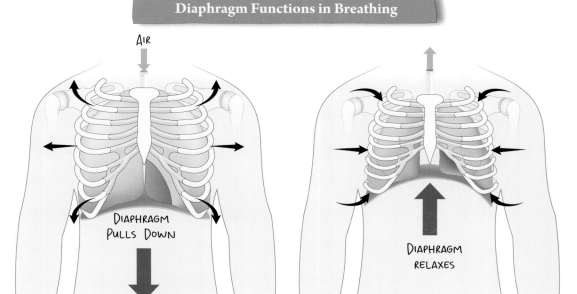

AIR

DIAPHRAGM PULLS DOWN

INHALATION

DIAPHRAGM RELAXES

EXHALATION

Inhaling and Exhaling Is Sophisticated: God prepared a very elaborate system to take oxygen out of the air and get it to every cell in our bodies for the many chemical reactions necessary for life. The same system then removes some of the waste (carbon dioxide) from every cell in our bodies. This is all done by the blood. The removal of carbon dioxide from the body occurs at the same time that the body receives oxygen: when the blood flows through lungs as air is inhaled and exhaled. This system works well in almost every human baby ever born.

How does a baby know to inhale and exhale? If this complex process does not operate several times per minute, he will die! A sensor determines that oxygen is needed. The brain tells the diaphragm (the large, dome-shaped muscle the lungs sit on) and rib-cage muscles to increase the size of the chest cavity and cause the lungs to draw in air. The lungs enlarge the right amount. Then the lungs shrink again, at the right speed. This rate of inhaling and exhaling is adjusted automatically to provide the necessary amount of oxygen for the body's activity rate.

IT IS A MARVELOUS SYSTEM THAT CAN BE CONTROLLED BY BOTH THE CONSCIOUS AND UNCONSCIOUS MIND. Most of the time, your unconscious mind makes you breathe. This is crucial since about one-third of your life you are asleep. However, God designed this system to allow your conscious mind to alter it: you can hold your breath when you go under water.

When oxygen is needed, a sensor is triggered and sends a signal to the brainstem (an "unconscious" part of the brain, sitting on top of the spinal cord, under the cerebrum). The brainstem then causes the dome-shaped diaphragm to flatten and the rib cage to expand. This makes a reduction in pressure, which enlarges the lungs by pulling in oxygen-rich air. This oxygen-rich air fills the baby's 50 million tiny air sacs in the lungs where the blood releases carbon dioxide and absorbs oxygen. This carbon dioxide-rich air is then pushed out of the body by the diaphragm relaxing back into its dome shape and the rib cage shrinking.[22]

The rate of inhaling and exhaling is adjusted automatically to provide the necessary amount of oxygen for the body's activity rate. Even newborn babies breathe as is necessary.

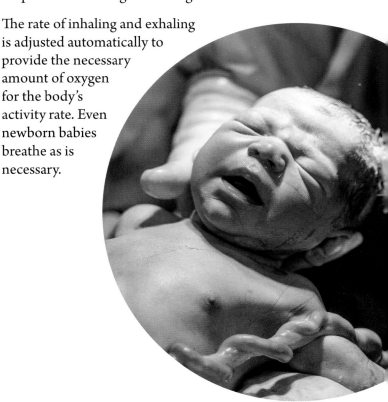

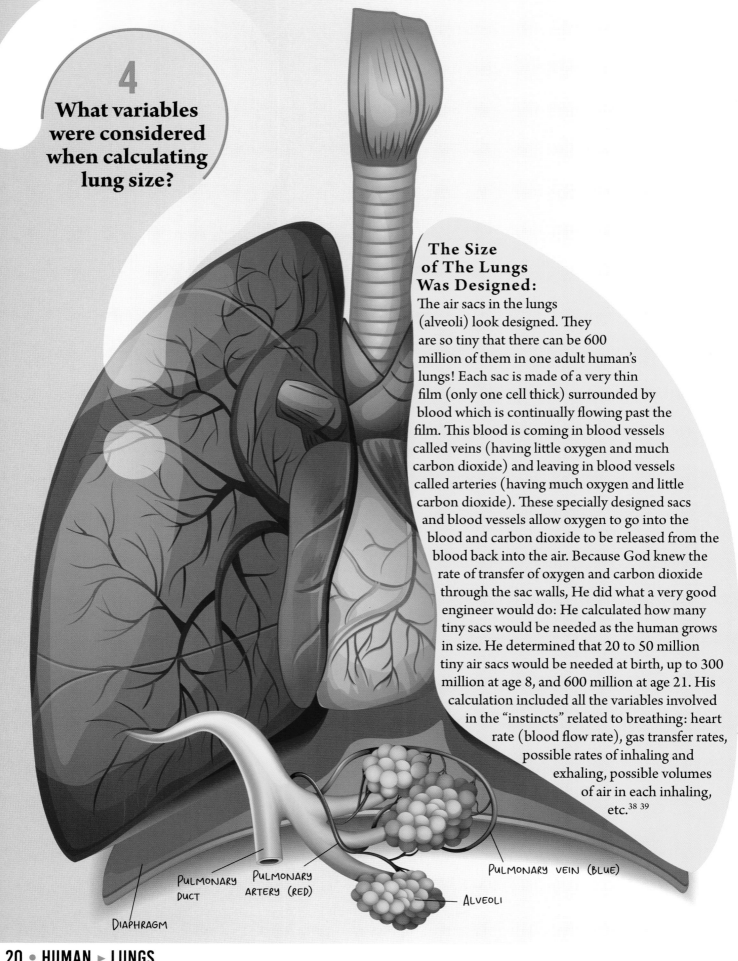

4
What variables were considered when calculating lung size?

The Size of The Lungs Was Designed:

The air sacs in the lungs (alveoli) look designed. They are so tiny that there can be 600 million of them in one adult human's lungs! Each sac is made of a very thin film (only one cell thick) surrounded by blood which is continually flowing past the film. This blood is coming in blood vessels called veins (having little oxygen and much carbon dioxide) and leaving in blood vessels called arteries (having much oxygen and little carbon dioxide). These specially designed sacs and blood vessels allow oxygen to go into the blood and carbon dioxide to be released from the blood back into the air. Because God knew the rate of transfer of oxygen and carbon dioxide through the sac walls, He did what a very good engineer would do: He calculated how many tiny sacs would be needed as the human grows in size. He determined that 20 to 50 million tiny air sacs would be needed at birth, up to 300 million at age 8, and 600 million at age 21. His calculation included all the variables involved in the "instincts" related to breathing: heart rate (blood flow rate), gas transfer rates, possible rates of inhaling and exhaling, possible volumes of air in each inhaling, etc.[38][39]

PULMONARY DUCT

PULMONARY ARTERY (RED)

PULMONARY VEIN (BLUE)

ALVEOLI

DIAPHRAGM

"Doctors didn't understand how we breathe until the beginning of the 20th century. Before then, when patients received injuries or had operations that opened their chests, they nearly always died of suffocation. Their lungs collapsed and would not reinflate."[23]

This is because the method of expanding the lungs is indirect. God put two almost empty balloons inside the human chest, one around each lung. When the chest is expanded, the balloons expand and pull on the lungs, making them "inhale." So, before the 20th century, surgeons would cut the balloon(s), and the lung(s) wouldn't work anymore.

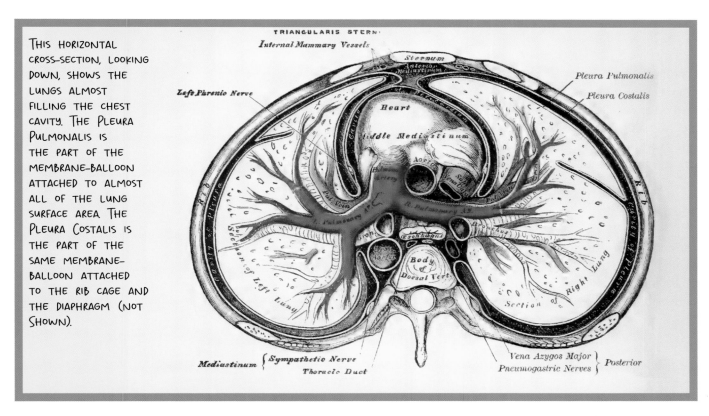

THIS HORIZONTAL CROSS-SECTION, LOOKING DOWN, SHOWS THE LUNGS ALMOST FILLING THE CHEST CAVITY. THE PLEURA PULMONALIS IS THE PART OF THE MEMBRANE-BALLOON ATTACHED TO ALMOST ALL OF THE LUNG SURFACE AREA. THE PLEURA COSTALIS IS THE PART OF THE SAME MEMBRANE-BALLOON ATTACHED TO THE RIB CAGE AND THE DIAPHRAGM (NOT SHOWN).

THE INSTINCTS BUILT INTO A BABY KNOW WHAT MUSCLES TO ACTIVATE FOR INHALATION AND EXHALATION.

The lungs fit very tightly inside the cavity created by the rib cage and the huge diaphragm muscle. God designed the rib cage with muscles that allow it to expand and contract. The diaphragm muscle is dome shaped and will flatten out when contracting. So this cavity is designed to grow and shrink, a perfect place for the lungs to expand and contract as you breathe.

In His design, however, instead of attaching the walls of the lungs to the diaphragm and the rib cage, God attached the airtight, almost empty balloons ("membranes") to the whole surface of each lung and to the inside of the rib cage and the top of the diaphragm muscle. Therefore, the lungs expand in every direction, not just toward the ribs and diaphragm. This method of expansion results in more air being pulled in than would happen if the lungs were directly attached to the rib cage and diaphragm.[24] It's a great design.

Why do we have two lungs instead of just one? Think about it. God provided duplicates of most of our bodies' crucial parts. If something goes wrong with one lung, we still would have the other to depend on. His planning is seen in even the fact that this balloon/membrane is not one, but two separate ones. That's why you can have "a collapsed lung" and still breathe — the other one still has a working balloon/membrane!

? What did surgeons finally learn in the twentieth century that made them stop collapsing lungs?

5
Why is it so important that we cough and sneeze?

Coughing: When food accidentally enters the TRACHEA (the windpipe) leading to the lungs, instead of entering the ESOPHAGUS leading to the stomach, even a baby's body instinctively reacts with what we call coughing.

If you were the designer of the human body, you would be very concerned about dealing with this mistake. If food goes into the trachea, the human could quickly die from lack of air reaching the lungs. As the designer, you might program the lungs to compress and blow out the blockage.[25]

God created a better way than that — one designed for a powerful blast of air. It includes muscles that are not used in normal breathing. This system (instinct) is used in those emergency situations of coughing and sneezing. As with most instincts, coughing is a complex set of instructions.

ESOPHAGUS

TRACHEA

Sneezing: Surely you have felt a sneeze coming on and tried to stop it. You probably failed. A powerful blast of air went through your nose, whether your conscious mind wanted it or not!

Sneezing is designed to clear the nasal area of something that should not be there. When a foreign body is detected by a baby's nose hairs, a signal is sent to the brainstem. The brain reacts with the sneezing instinct, a complex set of instructions. Just like the coughing instinct, these instructions organize a set of muscles to make a series of actions that result in a powerful blast of air.

SNEEZING

1	When something physical enters the trachea leading to the lungs, the lining of the airway sends signals to the brainstem.[26]
2	It causes the rib-cage muscles and the diaphragm to quickly expand the chest cavity to draw air into the lungs.
3	Once the inhalation is complete, the epiglottis seals the trachea (as it does when food is going down the other tube, the esophagus, toward the stomach).
4	With the epiglottis sealing the lungs, the brainstem makes the rib cage and diaphragm begin an exhalation, compressing the air.
5	After the planned amount of air pressure is achieved, the epiglottis opens the trachea.
6	The throat and mouth are opened and the powerful air blast goes through them. This should blow out almost any blockage.

The main difference between a cough and a sneeze is that the cough goes through the mouth, while the sneeze goes through the nose. This is done mainly by the brainstem making the tongue push against the roof of the mouth and seal it off, forcing all the air to go through the nose.[27]

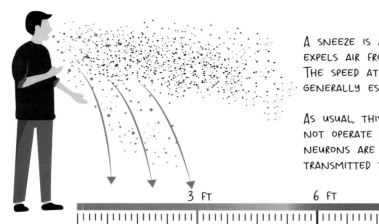

A SNEEZE IS AN INVOLUNTARY REFLEX THAT FORCEFULLY EXPELS AIR FROM THE LUNGS THROUGH THE NOSE AND MOUTH. THE SPEED AT WHICH A PERSON SNEEZES CAN VARY, BUT IT IS GENERALLY ESTIMATED TO BE AROUND 10 MILES PER HOUR

AS USUAL, THIS INSTINCT IS NOT SIMPLE. GOD MADE IT NOT OPERATE WHILE WE SLEEP. DURING SLEEP THE MOTOR NEURONS ARE NOT STIMULATED, AND REFLEX SIGNALS ARE NOT TRANSMITTED TO THE BRAIN.

3 FT   6 FT

MUCOCILIARY ESCALATOR

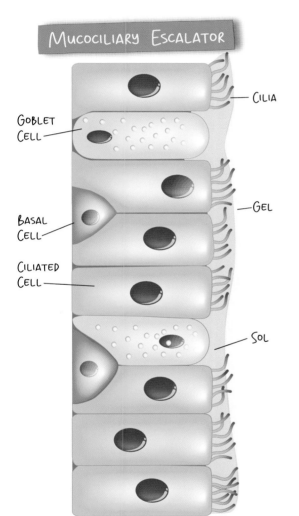

GOBLET CELL

BASAL CELL

CILIATED CELL

CILIA

GEL

SOL

Mucociliary Escalator: You've probably accidentally inhaled something, and coughing failed to get it all out. That particle may have been too small to be blown away from the wet wall of the trachea (windpipe). Later you realized that it was gone. It was probably removed by the Mucociliary Escalator.

God lined the trachea with very tiny hairs (cilia) coated in mucus. They are flexible, but stiff enough to always be pointing upward, away from the lungs. When foreign solids try to go down into the lungs, these wet hairs usually catch them. "Some of the important components of airway mucus are mucins (sticky, sugar-coated proteins), defense proteins, salt, and water. Together, these components form a gel that traps particles…"[28]

When the sensors inform the brainstem of the presence of the foreign solid, the instinct begins the process of moving it, with the mucus holding it, up and out of the windpipe. The method is to have the thousands of cilia move "the way your arms move while swimming the breaststroke."[29] An individual cilium is not very strong, but the instinct uses thousands of cilia "to produce coordinated, efficient movement of the mucus!"[30] Eventually, the foreign body and mucus will reach the top of the trachea and be removed by coughing or swallowing.

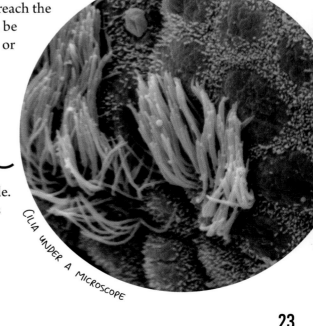

CILIA UNDER A MICROSCOPE

Coordinating thousands of hairs to move in a wave pattern is not simple. This led the writer of the above concerning cilia to make the ridiculous analogy of in a sense,

"the cilia talk to each other"[31] instead of giving God credit.

6

Why is the amount of water in your body so vital to your life?

Water is key to humans staying alive and healthy (it's true for most creatures God put on this planet). Consider what happens when you eat an apple. Over 1,300 chemical reactions occur as your body changes the chemicals in the apple into parts of your body.[32] All, or almost all, of those 1,300 reactions require water to happen. This is for just one apple. Aqueous (water-based) chemical reactions are continually happening all over your body. You need lots of water for all of those essential chemical reactions to occur.

Each of those many reactions inside your body requires reactants. Then, more reactants are needed for the next reactions. Quite often water is required for bringing new reactants and for taking away waste products. The main method of transport would be the continuously circulating blood, which is mostly water. Eventually the waste products that can't be recycled are removed from the body through sweat, urine, or solid waste (feces through the colon); all of which represent a significant loss of water.

The need for water is therefore obvious. However, the operation of our bodies is so complex that the correct amount of water is also key. There are several reasons for this. One reason for sensitivity to the amount of water is that our bodies use osmosis to get reactants inside places like cells for "numerous cellular functions" (the evolutionist writer actually said our bodies "exploited" osmosis; he gives our ignorant bodies credit for what God did).[33] What God did was build into our instincts the ability to use osmosis. When water is on both sides of a membrane (very thin film with very tiny holes), over time the concentrations of whatever tiny particles are in the water will become the same on both sides of the membrane as the particles randomly go through the membrane. If reactions using one kind of particle (a reactant) are going on continually on one side of the membrane (e.g., inside a cell), then that reactant will be continually flowing through the membrane into the cell replacing the used reactants. This is called osmosis. If the amount of water varies much, it could change the concentration of the reactant enough to affect the number of reactions inside the cells. Remember that there are many different reactions going on inside cells; most, if not all, are dependent on osmosis to get new reactants.

Another reason for sensitivity to the amount of water in the body is that the blood is mostly water. As the earlier

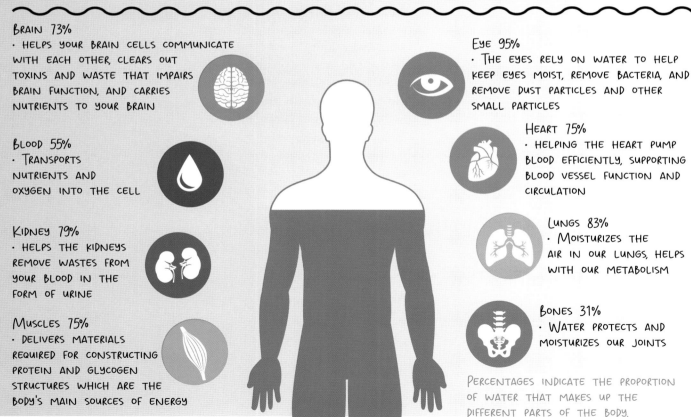

BRAIN 73%
· HELPS YOUR BRAIN CELLS COMMUNICATE WITH EACH OTHER, CLEARS OUT TOXINS AND WASTE THAT IMPAIRS BRAIN FUNCTION, AND CARRIES NUTRIENTS TO YOUR BRAIN

BLOOD 55%
· TRANSPORTS NUTRIENTS AND OXYGEN INTO THE CELL

KIDNEY 79%
· HELPS THE KIDNEYS REMOVE WASTES FROM YOUR BLOOD IN THE FORM OF URINE

MUSCLES 75%
· DELIVERS MATERIALS REQUIRED FOR CONSTRUCTING PROTEIN AND GLYCOGEN STRUCTURES WHICH ARE THE BODY'S MAIN SOURCES OF ENERGY

EYE 95%
· THE EYES RELY ON WATER TO HELP KEEP EYES MOIST, REMOVE BACTERIA, AND REMOVE DUST PARTICLES AND OTHER SMALL PARTICLES

HEART 75%
· HELPING THE HEART PUMP BLOOD EFFICIENTLY, SUPPORTING BLOOD VESSEL FUNCTION AND CIRCULATION

LUNGS 83%
· MOISTURIZES THE AIR IN OUR LUNGS, HELPS WITH OUR METABOLISM

BONES 31%
· WATER PROTECTS AND MOISTURIZES OUR JOINTS

PERCENTAGES INDICATE THE PROPORTION OF WATER THAT MAKES UP THE DIFFERENT PARTS OF THE BODY.

discussion of blood pressure showed, it is important to control blood pressure. One of the primary ways is to control the amount of water in the blood.

To show how amazing the thirst instinct is, please carefully read what an evolutionary scientist says about it. As you read, note how he repeats how the instinct's actions **must** happen for the sake of the body. Yet, he **believes** that they evolved (meaning over many thousands of years with no intelligence guiding it) even though so many steps **must** happen or the body suffers! He believes nothing designed it, as he admits he doesn't even know how it works (at the end of this quote he says the "how… remains unclear"). He has what can rightly be called blind faith in dumb luck!

Here is his description after saying fluid balance (amount of water in the body) is needed for the many chemical reactions and for correct blood pressure.

> "For these and other reasons, a physiologic system **has evolved** to maintain fluid balance. The key components of this system are **specialized** neurons that monitor the osmolality and volume of the blood and, when necessary, trigger a **coordinated set of autonomic**, **neuroendocrine**, and **behavioral responses** that defend these parameters against change. Many of these responses function at the level of the kidney, by modulating the rate at which water and salt are lost through excretion. However, there is a limit to the effectiveness of these renal **mechanisms, because** toxic substances **must** be periodically removed from the circulation through urination, and furthermore because sweating and other evaporative mechanisms cause continual loss of water and salt even in the absence of excretion. **Thus** at some point fluid balance **must** be restored by ingesting water.
>
> Thirst motivates water seeking and consumption by both positive and negative valence **mechanisms**. Thirst positively reinforces drinking behavior by magnifying the rewarding sensory properties of water: a glass of **water tastes wonderful when you are thirsty**, **but much less so when you are sated.** At the same time, thirst negatively reinforces drinking behavior by virtue of the fact that thirst itself is an unpleasant state, and thus animals are motivated to consume water in order to eliminate this aversive feeling. These **two motivational mechanisms act in unison** to promote drinking behavior, **but how** they are instantiated in the brain **remains unclear**"[34] (emphasis added).

Be sure and give God credit for what the evolutionist rightly saw: specialized neurons, a coordinated set of responses, renal mechanisms, periodic removal of toxic substances, restoration of fluid balance, positive and negative valence mechanisms, and all of it acting in unison. The instincts He created are amazing, even to an atheist!

SYMPTOMS OF DEHYDRATION

DRY MOUTH

THIRST

FATIGUE

DIZZINESS

DARK URINE

HEADACHE

WHEN YOU NEED MORE WATER

SPORTS OR EXERCISE

PREGNANCY

ENVIRONMENT

ILLNESS

7

How do newborns know they need to eat?

Have you felt hunger? The fact that even a baby knows he needs to get more energy and that the energy source (food) needs to go into a hole (the mouth) the baby HAS NEVER SEEN is beyond amazing! How does the baby know to put food into his mouth, rather than his ear or eye? Remarkably, when his body has received enough — he will stop being hungry.

"We couldn't have survived this long without the body's ability to feel hunger… (It) is a **very complex issue**."[35] Another report says the hunger instinct is a "mysterious function and… much remains to be learned."[36]

Humans of all ages have this important hunger instinct. Because adults often overeat to satisfy the hunger instinct, hunger will be discussed in more detail later on.

Human Babies' Desire to Eat: A brand-new human baby can do almost nothing. However, God has given him instincts about his need for the right food. His instincts tell him when he is hungry. This is a marvelous design that we usually do not appreciate. Even a newborn baby's body knows it is low in vitamins, minerals, energy, or liquid. When those sensors detect the need, they make the baby cry. Crying is an instinct designed to alert the mother or someone capable of supplying the baby's need. Crying will be discussed later.

The child's mother has instincts that quickly make her realize her child needs food. Because she recently gave birth to a child, her body has been creating food perfectly designed for the baby (milk), able to come out of the specially designed openings in her breasts. Those openings (nipples) are exactly the right size for the baby's sucking instinct to use.

Amazingly, even babies that have been out of the womb for only a few hours have the programming (instinct) to begin a nursing process, which is definitely not simple. For the previous nine months of his existence, he had NEVER EATEN anything; all nourishment had been brought by his blood exchanging nutrients and waste with his mother's blood. Now, suddenly that has changed. The baby does not consciously know it, but if he does not suck in milk soon, he will die!

With the mother's guidance and help, the baby begins to use his nursing instinct.

Scientists have named two reflexes involved in this nursing instinct of the baby.[37]	
1	First, the root reflex is the brain's reaction to anything brushing across his lips or cheek: he'll turn his head and open his mouth.
2	Second, the suck reflex gets the milk into the baby. When something touches the roof of his mouth, his instinct makes him start the sucking action. To actually obtain the milk, the lips must create an air-tight seal around the nipple, otherwise the rest of this complex set of actions would fail to feed the baby. Then air has to be pulled away from the lips by the lungs expanding (suction). This all works perfectly with the design of a mother's breast.

Instincts of Tasting and Feeding

1. When the milk is drawn into the mouth, the baby's tasting instinct begins. It quickly determines whether this liquid is good for the baby's body. If the tasting instinct says it is bad for the baby, the baby will force it all out of his mouth (spit).

2. Once the milk (or other liquid) is felt by sensors in the mouth and accepted by the tasting instinct, the swallowing instinct takes over. The expanding of the lungs is stopped. The tongue quickly pushes the milk (and very little air; as we grow older we learn to swallow almost no air) into the esophagus (the tube leading to the stomach).

3. Once that portion of milk is out of the way, the suction begins again for the next amount of milk. If the timing was not correct, the baby could easily suck milk into the windpipe, causing the need to cough. This swallowing instinct is so well designed that even babies rarely need to cough.

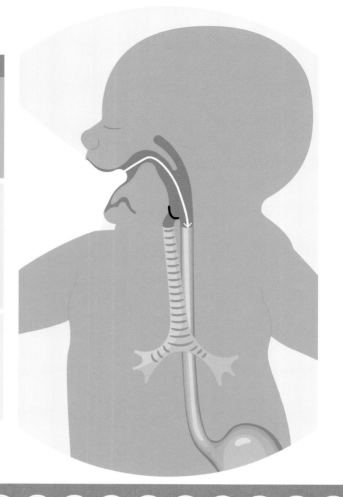

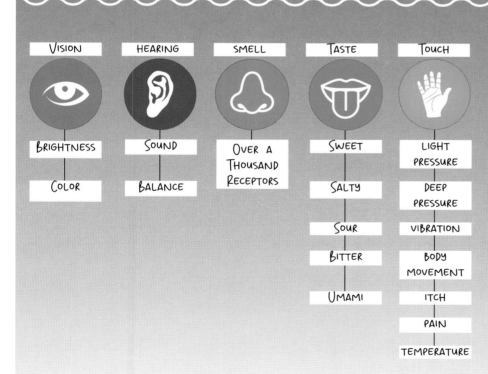

VISION	HEARING	SMELL	TASTE	TOUCH
BRIGHTNESS	SOUND	OVER A THOUSAND RECEPTORS	SWEET	LIGHT PRESSURE
COLOR	BALANCE		SALTY	DEEP PRESSURE
			SOUR	VIBRATION
			BITTER	BODY MOVEMENT
			UMAMI	ITCH
				PAIN
				TEMPERATURE

Are There Only 5 Senses?

People usually think of there being five senses: taste, touch, hearing, sight, and smell. There are actually at least 17! They are discussed in this book because almost all are used by our instincts to make our lives better.

To get 17 senses, the scientists will subdivide the five main ones. So, touch can be divided into light pressure, deep pressure, vibration, body movement, itch, pain, and temperature. [40]

God has provided many types of sensors and sense systems that are used by His built-in programming to make our lives better.

8 How does a mother's milk change to meet the needs of her baby?

What humans need to eat changes some as they get older, but the first months of a human life require unique food.

A baby needs the milk from his mother, because God has designed it to be the perfect food for him. The baby's instincts make him desire that milk more than anything else. Doctors today, unlike a few years ago, recommend that a baby breastfeed for the first six months, because mother's milk has everything a baby needs, including extra protections like giving immunities: "breast milk offers significant advantages over manufactured formula."[41]

Many instincts are involved in breastfeeding a baby. Instincts in the mother start preparing her body for breastfeeding while she is still pregnant. Once she gives birth, the milk starts to flow, but the amount of milk is more than one child would need (probably in case she had twins or triplets). The instinct quickly adjusts the amount to match what the baby or babies need.

The kind of milk is changed after the first few days from what is often called "liquid gold" (colostrum), because of what small amounts can do for the baby, to regular mother's milk. But regular mother's milk is "filled with special components that are **designed** to help fight infection…" (emphasis added to stress what the honest writer of the quoted material sees). Breastfed babies typically get less sick.[42]

Baby Food Makers Have a Dilemma

A baby should not eat much salt. As you get older you usually prefer a very salty taste. Why? Almost certainly because you've been eating American "food" which is loaded with salt and sugar. This is a problem for baby food manufacturers. They know the mother will test a food before giving it to her baby. They must choose whether to make food that is healthiest for the baby or tastes best to the mother! Unfortunately, they usually choose to make it taste better to the mother. [43]

Mothers tasting baby food need to remember that babies prefer a bland taste, much less salty than most mothers prefer.

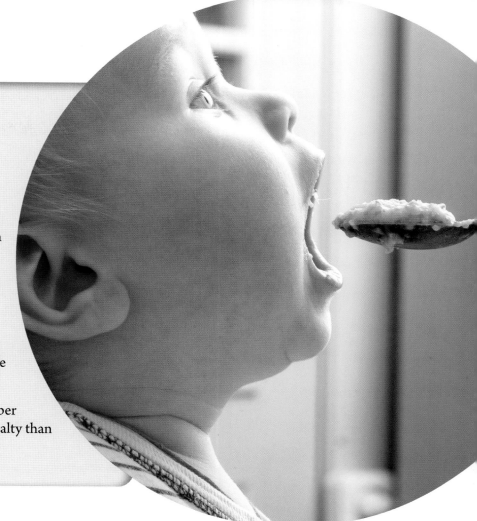

	The transition from milk to solid food is an important milestone in a baby's development. This process is commonly known as weaning. Here's a general overview of how a baby's body transitions from milk to solid food:	
1	Readiness	Babies are typically ready for solid foods around the age of 6 months, although individual readiness may vary. Key signs of readiness include the ability to sit up with support, good head control, and showing interest in food by watching others eat.
2	Tongue-thrust reflex	When babies are born, they have a natural instinct called the tongue-thrust reflex. This reflex causes them to push out anything placed on their tongue. As babies grow and develop, this reflex gradually diminishes, allowing them to begin swallowing and eating more solid foods.
3	Development of oral motor skills	As babies practice eating solid foods, their oral motor skills improve. They learn to move food around their mouth, chew with their gums, and swallow. Initially, the texture of the food should be smooth and soft, gradually progressing to thicker and lumpier consistencies.
4	Nutritional needs	Breast milk or formula continues to be an essential source of nutrition for babies during the transition to solid foods. As they consume more solids, their overall nutrient intake from complementary foods gradually increases, while their reliance on breast milk or formula decreases over time.

It's important to note that the process of transitioning to solid foods should be done gradually, based on the baby's cues and readiness. Every baby is different, and it's essential to consult with a healthcare provider for guidance and support during this transition.

9

In what way does your nose protect you from harm?

Desire for Good Taste: God has made most things that are good for humans to smell or taste good. Amazingly, most things that are bad for humans smell or taste bad.[44]

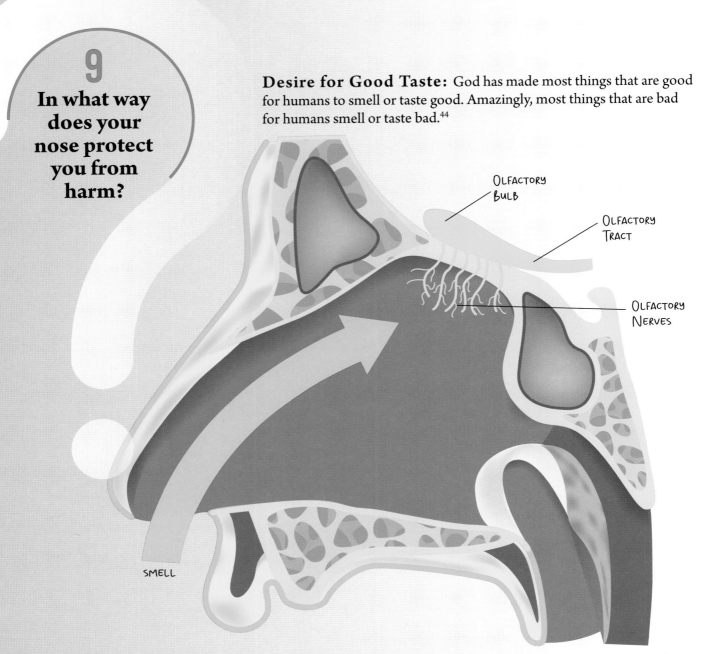

OLFACTORY BULB

OLFACTORY TRACT

OLFACTORY NERVES

SMELL

The Nose and Taste: Before food actually gets into your mouth, you usually smell it. Smell is the process where the millions of sensory neurons inside your nose detect molecules that have flown off of the "food" and into your nose. "Most scents are composed of many odorants; a whiff of chocolate, for example, is made up of hundreds of different odor molecules." They carry "complex information" that almost all humans can interpret.[45]

The sensors inside your nose, when the odor molecules bind to them, send nerve impulses to the thousands of clusters of neurons in the olfactory bulb in the

? About how many scents can humans detect with their noses?

cerebrum, which is the body's smell center. For each type of odor that enters the nose, a different set of neurons in the cerebrum are activated causing you to recognize the odor.

One scientist said, "Your nose has the **astonishing** ability to smell thousands of different scents…"[46] (emphasis added). He probably said this when scientists believed that we could detect only about 10,000 different odors. Imagine how astonished he was when scientists reported that the nose and brain are capable of distinguishing from each other roughly one trillion scents![47]

"But what about things that you know **smell good or bad even if you've never experienced them?** Scientists have found that although a lot of the smells people like come from past experiences, **instincts play a big role**"[48] (emphasis added). This statement is the point of this book: God has put into us many sets of information (instincts) that guide us (to healthy food) and protect us (from unhealthy things), even when the situation is new to us.

Biologists are trying to understand how we can know what is good or bad for us. "Our noses can quickly distinguish a pleasant smell and a stench, but until now the chemical cues that help us make such decisions had not been understood." The article proceeds to discuss how a scientist found that about 30% of the time molecular weight and electron density would indicate how a human nose would react to a smell. That means that 70% of the time his method did **not** explain how the human would react! This was followed by this admission:

"We think, **evolutionarily**, that our bodies have settled on this handful of properties to distinguish good smells from bad ones for a reason," Khan told LiveScience, but admitted that he's **uncertain why** the brain and nose **evolved** sensitivity to the molecular properties that they did.

"What we do know is that chemicals perceived as **less pleasant** are generally **not useful to us, and can even be harmful**," Khan said[49] (emphasis added).

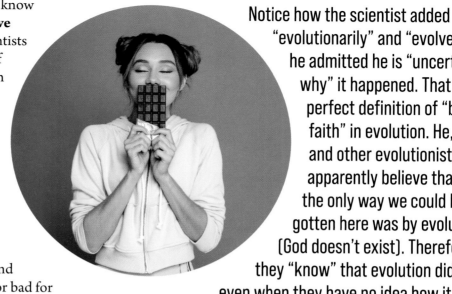

Notice how the scientist added "evolutionarily" and "evolved" as he admitted he is "uncertain why" it happened. That is a perfect definition of "blind faith" in evolution. He, and other evolutionists, apparently believe that the only way we could have gotten here was by evolution (God doesn't exist). Therefore, they "know" that evolution did it, even when they have no idea how it could have.

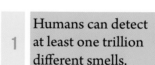

1	Humans can detect at least one trillion different smells.
2	Everyone smells slightly different.
3	Everyone's body smells slightly differently.
4	Women have a stronger sense of smell than men
5	Our sense of smell peaks during our late teens.

10

Why is it important that your tongue is the center of your taste receptors?

Taste consists of smell and what taste buds inside the mouth detect. Most TASTE BUDS are on the top and sides of the tongue, but some are above (on the soft palate) and behind it (on the pharynx). The microscopic taste buds (there are about 10,000 of them[50]) have dozens of taste receptor cells. The roughness you see on the tongue surface is apparently designed to determine the texture of what is in your mouth; the taste buds are on some of those bumps but are much smaller than the bumps.[51] Isn't it interesting that the taste buds surround the area where potential food enters your body, but is still where it can easily be expelled from your body before doing much harm (you can spit it out)?

The tongue may be able to detect a few dozen different sensations, but it works together with the nose to differentiate millions of flavors (a trillion of "scents" are considered "flavors"). The main sensations the tongue detects are sweet, bitter, sour, salty, and meaty (also called umami); some will include cool (mint) and hot (chilies).[52] However, the taste instinct is much more complicated than that: it can determine whether you should swallow something or not. This instinct is crucial for survival.

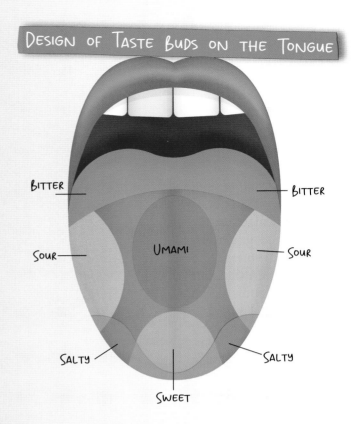

BITTER — BITTER
SOUR — UMAMI — SOUR
SALTY — SALTY
SWEET

Humans are omnivorous (eat all), rather than carnivorous (eat meat) or herbivorous (eat plants). So, our menu of foods is more varied than many animals. "Taste is an **especially important** sense for omnivorous species given that the potential range of foods, their **variation in nutrient content**, and the **hazards of accidental toxin ingestion** increase with the variety and complexity of the feeding strategy"[53] (emphasis added). Animals with a limited selection of food will have less chance of eating something bad for them.

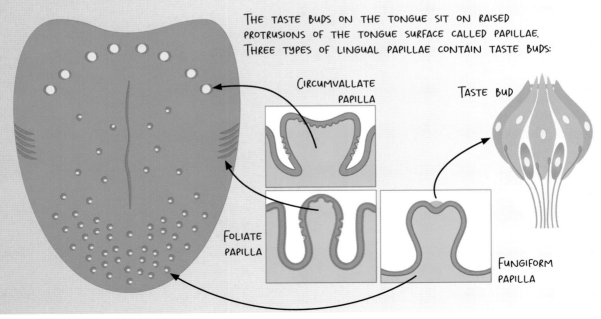

THE TASTE BUDS ON THE TONGUE SIT ON RAISED PROTRUSIONS OF THE TONGUE SURFACE CALLED PAPILLAE. THREE TYPES OF LINGUAL PAPILLAE CONTAIN TASTE BUDS:

CIRCUMVALLATE PAPILLA

TASTE BUD

FOLIATE PAPILLA

FUNGIFORM PAPILLA

Imagine being lost for days out in the woods. How would you know what to eat? The vast majority of the time, your taste instinct would correctly determine whether you should swallow or not. Look at the main sensations of taste and what they generally represent in things people could put into their mouths:

1. Sweet foods tend to be full of necessary calories and carbohydrates, so our taste instinct makes us like them.

2. Bitter foods tend to be toxic or harmful, so the instinct says to avoid them.

3. Sour foods also tend to be toxic or harmful, and the instinct says to avoid most of them.

4. Salty foods are rich in the important electrolyte salt, which helps our bodies function properly, so our instinct encourages us to eat them.

5. Meaty foods tend to have a lot of amino acids and proteins which are important for body parts such as muscles, so the taste instinct encourages us to eat them.[54]

WHEN YOUNG, THE FUNGUS IS EDIBLE AND TASTES LIKE CHICKEN, HENCE THE COMMON NAME "CHICKEN FUNGUS"

Remember that the brain considers signals from the tongue and from the nose as it encourages or warns us about things we are considering eating. How would a child have all of these very important pieces of information built into him?

Because a scientist believed evolution slowly created our taste instinct, he said that when we eat something new to us, "If the outcome is positive, taste will signal pleasure and reward…."[55] Think about it. He seems to be saying that not only will we remember that that item was good to eat, but *somehow* it will be incorporated into our DNA and be passed down to future generations as part of the taste instinct. That was what Lamarck taught in the early 1800s … until it was disproven by real science!

This same scientist's descriptions are so unclear (probably because he has no clear idea of how the vital taste instinct "evolved") that one could easily conclude he thinks the earliest creatures had a full set of taste instinct "controls," but the unused ones disappeared in future evolved creatures.[56] So, creatures with very limited diets have fewer sensors, because the many other sensors were lost over the generations due to lack of use. This is the opposite of what evolution requires: this is complex to simple, rather than simple to complex. God creating every creature to reproduce after its own kind, with all of the necessary instincts for that kind, fits the data much better.

TO MAKE IT CLEARER THAT THERE HAD TO BE A DESIGNER, realize humans are not the only creatures that know what to eat or not eat. Almost all, probably all, living things on this planet know what they should eat and not eat! And they generally don't eat the same things. The things that make you throw up are the favorite food of other creatures!

11
How does a child know how to chew and swallow food?

Once your instincts have decided that something is good to eat another set of instincts come into play.

Overview of Digestion: "Your digestive system is **uniquely constructed** to do its job of turning your food into the nutrients and energy you need to survive. And when it's done with that, it **handily packages** your solid waste, or stool, for disposal when you have a bowel movement"[57] (emphasis added).

This is a great summary of what God designed for all humans. Notice that the emphasized words would naturally mean a Designer, but, as usual, the authors don't give Him any credit. If they don't believe in Him, they certainly still see design (Romans 1:18–20). If they do believe in Him, they are probably afraid their work won't be published if they state their belief.

> "The main organs that make up the digestive system (in order of their function) are the mouth, esophagus, stomach, small intestine, large intestine, rectum and anus. Helping them along the way are the pancreas, gall bladder and liver."[58]

Baby Upper Teeth	
Central incisor	8–12 months
Lateral incisor	9–13 months
Canine	16–22 months
1st Molar	13–19 months
2nd Molar	25–33 months

Baby Lower Teeth	
2nd Molar	25–31 months
1st Molar	14–18 months
Canine	17–23 months
Lateral incisor	10–16 months
Central incisor	6–10 months

Biting and Chewing Food: The first piece of the digestive system is the mouth. A baby is not born with teeth, but as soon as they begin to appear in his mouth, he knows how to use them. Even baby teeth are designed to cut, tear, and grind food. The mouth is designed to have the cutting teeth in the very front (the incisors). Next to them are our tearing teeth for what does not cut easily (the canines). In the back are the teeth that grind down what you are able to cut or tear or what just needs size reduction (the molars). Those nicely designed teeth will do nothing unless the jaw muscles and the tongue move food correctly to cut, tear, or grind it. The tongue maneuvers the food to the correct set of teeth (for cutting, tearing, or grinding). How does a child, who has probably never seen inside his mouth, know where to push the food for the next step in size reduction?

The child will instinctively continue to chew the food until it is small enough to swallow. How does he know how small it needs to be?

Once the teeth finish their work, the tongue guides the softened pieces to sit on the middle of the tongue. How does a baby know that chewed food needs to sit on the tongue in order to be "swallowed"?

The answers to these questions are that the complex programming (instinct) guides the baby's brain. The instinct is important, because the mouth is designed to be a part of the digestive system. It does the initial breaking down of a piece of food into much smaller pieces with the teeth.

The **Mouth** is the first part of the upper gastrointestinal tract. A mechanical chewing of food and wetting of food with saliva occurs in the mouth. Your **Tongue** is thick and mostly made of muscles that aid in chewing, speaking, and breathing. Your tongue runs from the middle of your neck to the floor of your mouth.

SALIVA (you have probably called it spit) is key to preparing the pieces of food for digestion.

Many glands positioned appropriately all around the teeth, gums, and tongue produce the liquid called saliva. Saliva is designed to lubricate the food pieces and to start the digestion. Saliva is 98% water but it "contains many important substances, including electrolytes, mucus, antibacterial compounds and various enzymes."[59] The water lubricates the food pieces and helps create soft food balls that can be swallowed easily. The enzymes perform tasks such as starting the breakdown of complex carbohydrates into simple sugars, ready to be used as energy in the body.[60]

Have you ever thought about how complex SWALLOWING is? Actually, you probably have. The first time you tried to swallow a large capsule of medicine, you probably struggled with how to get that big thing down your throat. You may have failed and gagged. Eventually your cerebrum (the "conscious" part of the brain) figured it out and you swallowed the medicine, probably as if it were a small amount of water on the middle of your tongue. However, when you ate food, even as a baby, you knew how to swallow! Scientists are fairly certain that the brainstem (the unconscious part of the brain) controls swallowing.[61]

This explains how a newborn baby can swallow: God put the program into the baby's brainstem. This book will call that program the swallowing instinct.

Swallowing is a part of the human digestive system. After your instincts of biting and chewing have prepared food for digestion in the stomach, it must be moved down to the stomach; it must be "swallowed."

While the food pieces sit on the tongue, the saliva glands will flood the front of the mouth. When enough saliva is in the front of the mouth, the tongue, cheek muscles, and back of the mouth muscles will push the food pieces into the now open throat. If you look at a side-view drawing of the inside of the mouth, you will probably be surprised by what the muscles look like, such as how thick your tongue is. So how can a baby know how to maneuver food through that passageway with that set of muscles to get to the entrance of the ESOPHAGUS (the tube connecting the throat and the stomach)?

Note that only his instincts can prevent food from going the wrong way. From the mouth there are three directions food could go, and two of the three would cause problems. The instincts (the unconscious mind) routinely handle this with little thought from your conscious mind. They cover the opening to the nose with the soft palate.[62] They cover the opening into the lungs with the epiglottis.[63]

The esophagus, the third and correct path for food from the mouth, is lubricated and has a sphincter muscle to open and close it whenever the baby's instincts command it. Once the food enters the esophagus, this amazing instinct makes the esophagus muscle keep contracting and relaxing until the food has been pushed all the way down into the stomach.

Thus, your conscious mind starts the process of swallowing: putting food into your mouth and beginning to chew. Your instinct takes over with the basics of chewing and when to push the food toward the throat, though the conscious mind could still control those steps. Once the food is pushed into the esophagus, only the instinct has control.

IT IS ANOTHER AMAZING, COMPLEX PROGRAM OF INSTRUCTIONS, NECESSARY FOR LIFE TO CONTINUE; EVEN BABIES HAVE IT.

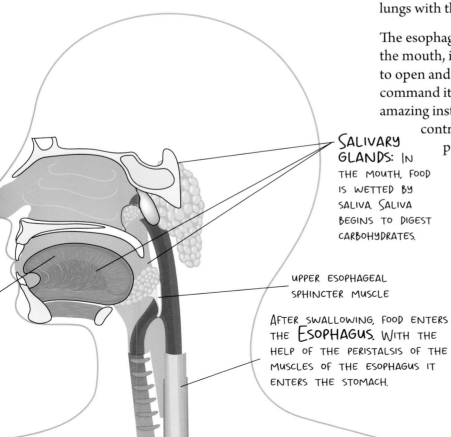

SALIVARY GLANDS: IN THE MOUTH, FOOD IS WETTED BY SALIVA. SALIVA BEGINS TO DIGEST CARBOHYDRATES.

UPPER ESOPHAGEAL SPHINCTER MUSCLE

AFTER SWALLOWING, FOOD ENTERS THE ESOPHAGUS. WITH THE HELP OF THE PERISTALSIS OF THE MUSCLES OF THE ESOPHAGUS IT ENTERS THE STOMACH.

12

When does your stomach know to start metering out digestion solution?

The ESOPHAGUS is a special tube for taking food from the mouth down to the stomach. This tube is one big muscle that is able to squeeze in waves (pulses). After food enters the tube, a circular muscle (called a sphincter muscle) closes the opening behind it. Stretching caused by the food will stimulate the nerves in the esophagus and start the rhythmic pulsing. The muscle making up the esophagus tube begins contracting and relaxing in a well-ordered method controlled by instinct. It contracts at the top, pushing the lump of food down toward the stomach. Before the tube begins to relax at the top, the section now holding the food begins to contract, pushing the food farther down. Before the second section begins to relax, the section now containing the food begins to contract.

It is very much like you squeezing tightly one end of a soft straw and sliding your tightly-pressing fingers down the length of the straw; any liquid in the straw would be pushed out the far end of the straw. This contracting and relaxing moves the food down the esophagus until it reaches the stomach.[64] But unlike the straw, the bottom of the esophagus needs to stay closed as much as possible to keep stomach acid out of the esophagus. The instinct controls the timing of this.

This design is so good that it is used in many pieces of equipment by engineers. They call it a peristaltic pump, and they call the process positive displacement. God designed it first and put it into use in very necessary places like the esophagus, the stomach, and the intestines.[65]

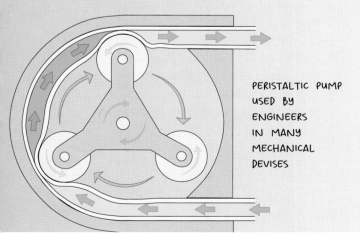

PERISTALTIC PUMP USED BY ENGINEERS IN MANY MECHANICAL DEVISES

How did God design the esophagus to keep the stomach's strong acid from going into the mouth while a child is hanging upside down or an astronaut is floating in zero gravity in a space craft? He put a SPHINCTER MUSCLE at the bottom of the esophagus. Sphincter muscles are used in several places throughout the human body. When they are relaxed, they are shaped like donuts — a ring with a hole in the middle. When the sphincter muscle contracts, the hole in the middle is completely sealed shut. People whose esophagus bottom sphincter muscle has stopped completely sealing off the stomach acid have what is called acid reflux or heartburn.[66]

The STOMACH is a balloon made of muscle. If you haven't eaten in many hours, it is probably almost flat ("empty"). A full, adult stomach will be about the size of a 2-liter bottle of soda (2.3L is the average maximum capacity).[67, 68] It does not get full only from what you eat. For the food you put into your mouth, the instinct will put about the same amount of acid into the stomach.[69]

When the vagus nerve senses that food has entered the stomach, the brainstem starts the set of instructions for digesting it (instinct). The mouth's saliva starts the digestion of the food, but the stomach does the majority of the breaking down of food into small, useful molecules. It does this both chemically and mechanically.

The chemical digesting is started by 3 types of cells in the lining of the stomach secreting acid and powerful enzymes.

1	One type of cell starts making strong hydrochloric acid (about pH=2), which will dissolve most food.
2	A second type of cell begins producing an inactive enzyme (pepsin) that needs the hydrochloric acid to become active and do the difficult job of digesting proteins.
3	The third type of cell produces mucus to coat the stomach lining to better protect it from the strong acid.

The mechanical digesting is done by the whole stomach pulsing. This large muscle is able to churn the solution of food and chemicals inside it. This mixing is improved by the fact that the stomach lining has ridges (a smooth lining would produce less turbulence). This mixing action goes on for about 5 hours (the actual time is controlled by the instinct).

At the right time, the instinct's instructions will have the sphincter muscle at the bottom of the stomach start metering out (regulating and releasing in a controlled manner) the digestive juices and enzymes, called "chyme" at a slow, controlled rate into the SMALL INTESTINE.[70]

AFTER SWALLOWING, FOOD ENTERS THE ESOPHAGUS. AND IS MOVED TOWARD THE STOMACH BY THE PERISTALTIC CONTRACTIONS OF THE ESOPHAGEAL MUSCLES.

LOWER ESOPHAGEAL SPHINCTER MUSCLE

PYLORIC SPHINCTER MUSCLE

THE STOMACH IS THE ORGAN WHERE THE DIGESTION OF PROTEIN OCCURS, USING GASTRIC ACID AND OTHER DIGESTIVE JUICES.

This is probably much more complicated than you thought. Realize that you normally do not feel much if any of all this action. However, if something goes wrong, you can feel so much that you call it a "stomach ache!" These instincts are amazing.

After about 5 hours in the stomach (depending on how much and what kind of food it is), the instinct will send the digesting food on from the stomach into the small intestine.[71] The main actions of the small intestine could be summarized as finishing the digestion of the food and removing the nutrients and water. The method is by adding more water and chemicals at the beginning, churning it while pushing it forward, and slowly removing most of the nutrients and water. The instinct controlling this process is very complex.

The small intestine in an adult human is a 10-foot-long, 1-inch-diameter tubular muscle[72] (it's 22 feet long when the muscle completely relaxes[73]) similar to the esophagus in that it is a peristaltic pump. However, it is like the stomach in that it adds more chemicals to continue the digestion of the food. It is different in that it begins the important process of removal of nutrients and water.

When the very viscous solution (chyme) enters the small intestine, the instinct causes the pancreas to send enzymes for breaking down proteins, fat, and carbohydrates. The liver is ordered to send enzymes for digesting fat and some vitamins.

The churning and addition of water in the first third of the small intestine finishes reducing the chyme to a liquid similar to water in viscosity. The rest of the small intestine is mainly for removing the nutrients and water as the liquid food is pushed along its length.

IN THE SMALL INTESTINE, DIGESTION OF CARBOHYDRATES OCCURS, INCLUDING FATS, POLYPEPTIDES, NUCLEIC ACIDS AND ABSORPTION OF THE AMINO ACIDS, PEPTIDES, GLUCOSE, FRUCTOSE, AND FATS.

(?) Since your stomach is full of acid, how are kids able to hang upside down without acid going up into their mouths?

13

How many chemical reactions occur in your body every second?

Since the rate of removal of nutrients depends on the area of contact between the liquid and the inside wall of the small intestine, God used an ingenious design. Instead of a smooth surface, He made it rough. The rough design is described as folds (villi) which are each full of tiny folds (microvilli). See the figure below. The interior surface area of this 22-foot tube for removal of nutrients and water is actually the size of a tennis court![74] The food solution will normally take about 5 hours to go through an adult small intestine.[75]

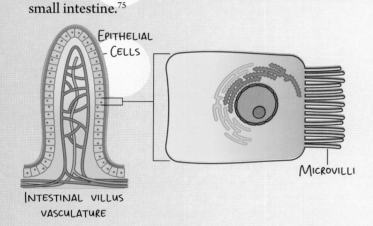

EPITHELIAL CELLS

MICROVILLI

INTESTINAL VILLUS VASCULATURE

A major job of the small intestine is to remove nutrients from the food you eat. However, without the LIVER and the PANCREAS the system would barely work. The liver produces bile that breaks down fats and the pancreas produces several enzymes that break down protein, starch, and fats. The pancreas also reduces the acidity as digestion is finishing. As usual, this is a complex system.

The instincts monitor when bile and the enzymes are needed in the small intestine. Bile is continually being made by the liver, but fat in the small intestine stimulates the hormones to produce more. When the small intestine is empty, the duct leading to it from the liver is closed by the instinct, diverting the new bile to the GALLBLADDER where it is stored and concentrated for later use.

The production of enzymes in the pancreas is marvelous.

"The pancreas produces protein-digesting enzymes in their **inactive forms.** These enzymes are activated in the duodenum (the beginning of the small intestine). If produced in an active form, **they would digest the pancreas** (which is exactly what occurs in the disease, pancreatitis)….

The enzymes that digest starch (amylase), fat (lipase), and nucleic acids (nuclease) are secreted in their **active forms, since they do not attack the pancreas** as do the protein-digesting enzymes..."[76] (emphasis added).

Note from this previous quote how the writer does not seem to be amazed at all that the pancreas "knows" to make some enzymes in their inactive form, but others are okay to be made in their active form. This doesn't even mention the fact that there must be something in the small intestine to activate the inactive ones (the enzyme enteropeptidase does the activating, and, of course, it "happens" to be in the right place, waiting).

All of the above (and more) are needed to get the nutrients out of your food. What happens to the nutrients removed from the food? THE INSTINCT HAS THE LIVER TAKE THESE "RAW MATERIALS" AND CREATE "ALL THE VARIOUS CHEMICALS THE BODY NEEDS TO FUNCTION."[77]

But it's more complicated than that. What about the bad things you accidentally eat? Almost everything extracted by the intestines is sent to the liver. It has many different kinds of cells which are designed to handle the good and the bad of what you eat. **"By cooperating,** they can filter the blood, store vitamins and minerals, excrete harmful toxins, produce bile, transport materials, form compounds that help coagulate the blood and metabolize carbohydrates, fats and proteins"[78] (emphasis added). Notice again that "cooperation" requires some kind of programming by a Programmer.

Any chemist or biologist would have a very difficult time making a few of these many chemicals that a baby's body does several times a day! If the baby's body fails to make even some of these chemicals, he will get sick and probably die within a short time. GOD KNEW EVERY CHEMICAL, EVERY ELEMENT WE WOULD NEED TO LIVE AND THRIVE, AND HE MADE INSTINCTS TO CREATE THEM WHEN AND WHERE THEY ARE NEEDED!

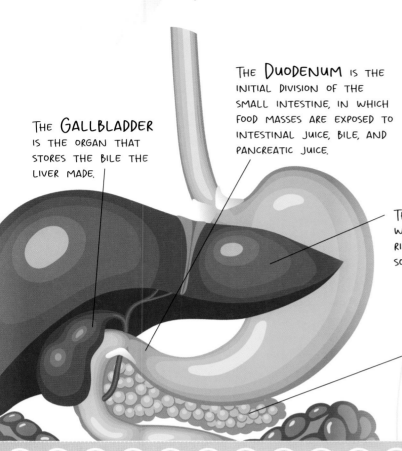

THE GALLBLADDER IS THE ORGAN THAT STORES THE BILE THE LIVER MADE.

THE DUODENUM IS THE INITIAL DIVISION OF THE SMALL INTESTINE, IN WHICH FOOD MASSES ARE EXPOSED TO INTESTINAL JUICE, BILE, AND PANCREATIC JUICE.

THE LIVER IS THE ORGAN THAT PRODUCES BILE, WHICH IS USED TO DIGEST FATS. THE LIVER GETS RID OF WASTES IN THE BODY, CLEANS ALCOHOL, SOME MEDICINE, AND POISONS FROM THE BLOOD.

THE PANCREAS IS THE GLAND THAT MAKES ENZYMES FOR DIGESTION AND THE HORMONE INSULIN.

(?) When you eat, what really happens to the food? About how long does it take to go all the way through you?

Your Body Is an Unbelievable Chemist!

You've been told to drink water for good health. Water is necessary for many functions in your body. It is necessary for converting nutrients into things like body parts. As this book is trying to stress to you, the complexity of life is often more than you had imagined, but what water is used for is beyond your comprehension!

Almost all chemical reactions in the human body are aqueous (require water to happen). A good estimate of the number of cells in your body is about 37 trillion. Biologists have determined that each of these "simple" cells in your body uses water to perform about one billion reactions **per second**! That means that as you read this your body is performing the astronomical number of 37 thousand billion billion reactions every second!

How could so many reactions be happening in each tiny cell? The keys for any reaction are distance between reactants and their speed (which is a function of temperature). "**It just so happens** that the average Eukaryotic cell, of which all multicellular organisms are comprised, has a volume of about 1 x 10-11 liters, which **nature selected** over time because it is **the perfect size for the most efficient biochemistry,** i.e., the **perfect** number of collisions per unit time **at body temperature,** which is 37 degrees Celsius for humans"(emphasis added).[79]

Notice that this biologist saw the design was "perfect," but as Romans 1:18–20 says, he would not give God credit. He said that "nature selected over time." Time seems to be his god. Common sense says that if evolution were true and you search back in time you'll find "imperfect" sizes and temperatures. They don't find any significant changes "over time." If you're open to hearing God, it is obvious that He designed all of this!

14

Why is the shape of the rectum's interior so vital for eliminating waste?

After almost all of the nutrients and water have been removed in the small intestine, the instinct sends the now viscous liquid into the LARGE INTESTINE (it is also called the colon, except for the short first section). The main job of this 3-inch-diameter, 6-foot tubular muscle is to remove more water, but it will also remove most of the remaining nutrients, sending them to the liver.

This absorption of water and nutrients in both intestines is not simple; the instinct controls it with "neuroendocrine mechanisms."[80] As the mass moves through the large intestine it becomes more and more solid, though at the end it is still about 75% water.[81] The final concentration depends on many factors: what was eaten and the time in the large intestine are the primary factors. Researchers have determined that the travel time through the large intestine varies from 10 to 60 hours, with the average being about one and a half days.[82]

How can it take so long in the large intestine while it's being fed by the small intestine where it only stays about 5 hours? Simple math will show that reducing the water content from about 98% to about 75% will reduce the total volume to less than one tenth! Plus, going from 1-inch diameter to 3-inch diameter will increase the large intestine cross-sectional area to 9 times as much as the small intestine. The working lengths are very similar: 10 feet for the small intestine and 6 feet for the large intestine. That's why the time in the small intestine can be about 5 hours while the time in the large intestine can be up to 60 hours.

While in the large intestine the waste solution is being churned continuously (to make more efficient the removal of water and the preparation of the waste solids for expulsion from the body), but it is not moving along the tube at all. Only once or twice per day the brainstem's programming (instinct) determines another aliquot of waste is ready to be expelled from the body. The instinct changes the pulsing of the large intestine. The pulsing becomes a uniform peristaltic contracting. This moves the semi-solid waste (feces) to fill the 8-inch-long rectum, which is at the end of the colon and normally empty.[83]

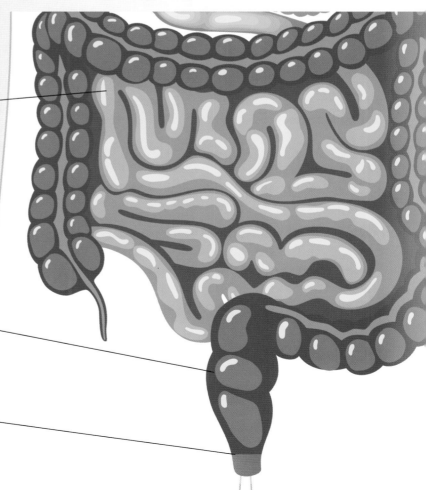

THE LARGE INTESTINE CREATES FERMENTATION OF THE FOOD MASSES BY INTESTINAL FLORA, SYNTHESIS OF VITAMINS, AND ABSORPTION OF WATER AND GLUCOSE, FRUCTOSE, VITAMINS AND MINERALS. THE LARGE INTESTINE FORMS SOLID WASTE FROM DIGESTION.

THE RECTUM IS THE LOWER END OF THE LARGE INTESTINE, LEADING TO THE ANUS. SOLID WASTE PRODUCTS FROM DIGESTION MOVE INTO THE COLON AND LEAVE THE BODY VIA THE RECTUM.

THE ANUS IS THE OPENING AT THE FAR END OF THE DIGESTIVE TRACT THROUGH WHICH SOLID WASTE PRODUCTS FROM DIGESTION LEAVE THE BODY.

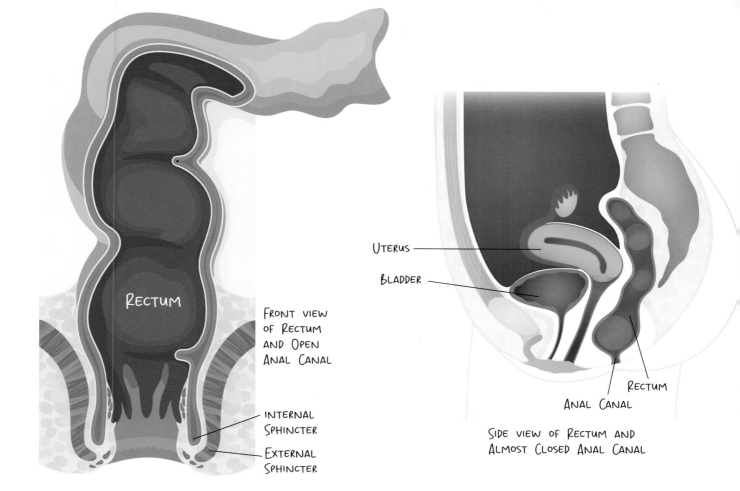

FRONT VIEW OF RECTUM AND OPEN ANAL CANAL

RECTUM

INTERNAL SPHINCTER

EXTERNAL SPHINCTER

UTERUS

BLADDER

RECTUM

ANAL CANAL

SIDE VIEW OF RECTUM AND ALMOST CLOSED ANAL CANAL

Rectum and Anal Canal: The design of the end of the large intestine is seen by even atheists to have several purposes. There is a 90-degree bend with an internal ridge which makes the last 8 inches of the large intestine able to stay empty until the instinct commands its filling. Doctors call this section the rectum. *Rectum* is Latin for straight, and the outside looks straight. However, researchers have determined that the interior is not straight and is well designed.

The shape of the interior of the rectum forces the waste solids (feces) to move left and right and forward and backward. The only exit is at a right angle into the anal canal. See the two figures above. Studies have shown several reasons for all of this bending and turning. It tends to separate the gas from the solids (this would reduce explosive events on the toilet).[84] It also reduces the gravitational pressure on the anal canal, thus it's easier for the human's conscious mind to control when waste will be expelled.[85]

The anal canal is just a 1-inch-diameter, 2-inch-long opening to the outside of the body. The interior inch and a half is completely closed by the brainstem-controlled (involuntary) internal sphincter muscle. The exterior half inch is completely closed by the cerebrum-controlled (voluntary) external sphincter muscle.

When waste enters the rectum, sensors signal the brainstem. The instinct determines how much gas and solid is there and whether it is ready to be expelled. If it is to be expelled, the brainstem causes the internal anal sphincter muscle to open. Most importantly, this instinct warns the conscious mind that the waste is ready to be removed.[86]

A baby generally doesn't know how to control the external anal sphincter muscle, so the waste will immediately come out into his diaper. The adult will have control over that external sphincter muscle. The instinct is complex enough to allow the conscious mind to delay expelling the waste; the signaling to the cerebrum will temporarily stop.[87] When the adult conscious mind opens the external anal sphincter muscle, the brainstem will signal the muscles of the rectum to start pushing.

15

In what way does the instinct for adding ADH at night help you rest?

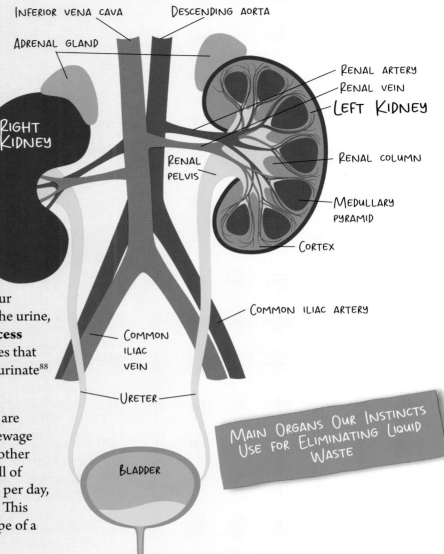

INFERIOR VENA CAVA

DESCENDING AORTA

ADRENAL GLAND

RIGHT KIDNEY

RENAL ARTERY
RENAL VEIN
LEFT KIDNEY
RENAL COLUMN

RENAL PELVIS

MEDULLARY PYRAMID

CORTEX

COMMON ILIAC ARTERY

COMMON ILIAC VEIN

URETER

BLADDER

MAIN ORGANS OUR INSTINCTS USE FOR ELIMINATING LIQUID WASTE

Urinating: The elimination of the liquid waste (called urine) is very important. You might think it is simply your mind detecting that the bladder, holding the urine, is full. "It's actually **a pretty complex process** involving many muscles, organs, and nerves that work together to tell you that it's time" to urinate[88] (emphasis added).

The main organs of the URINARY SYSTEM are the two KIDNEYS. They are your body's "sewage treatment plant" for removing toxins and other substances no longer needed. They filter all of your blood (about 7 liters) over 300 times per day, removing 1½ to 2 liters of urine each day. This urine is sent to a hollow muscle in the shape of a balloon (the "bladder").[89]

In a baby, the urinating instinct will do everything. It will open the bladder sphincter, make the bladder squeeze out the urine, and open the external sphincter and fill the baby's diaper with urine. As the baby grows up, he learns to control that external sphincter muscle and to suppress the messages from the bladder nerves. In adults, all that the conscious mind can control is the external sphincter muscle, as it fights the message from the bladder that it needs to be emptied.[92]

"Although **the most important bodily functions work right after birth, the fine-tuning of the organs takes time.** This also applies to bladder control, which takes longer to develop in some children and can't be sped up"[93] (emphasis added).

This quote shows how that scientist realized that God, through instincts, gives even newborns the "most important bodily functions," yet allows the conscious mind to eventually have some control over many of them.

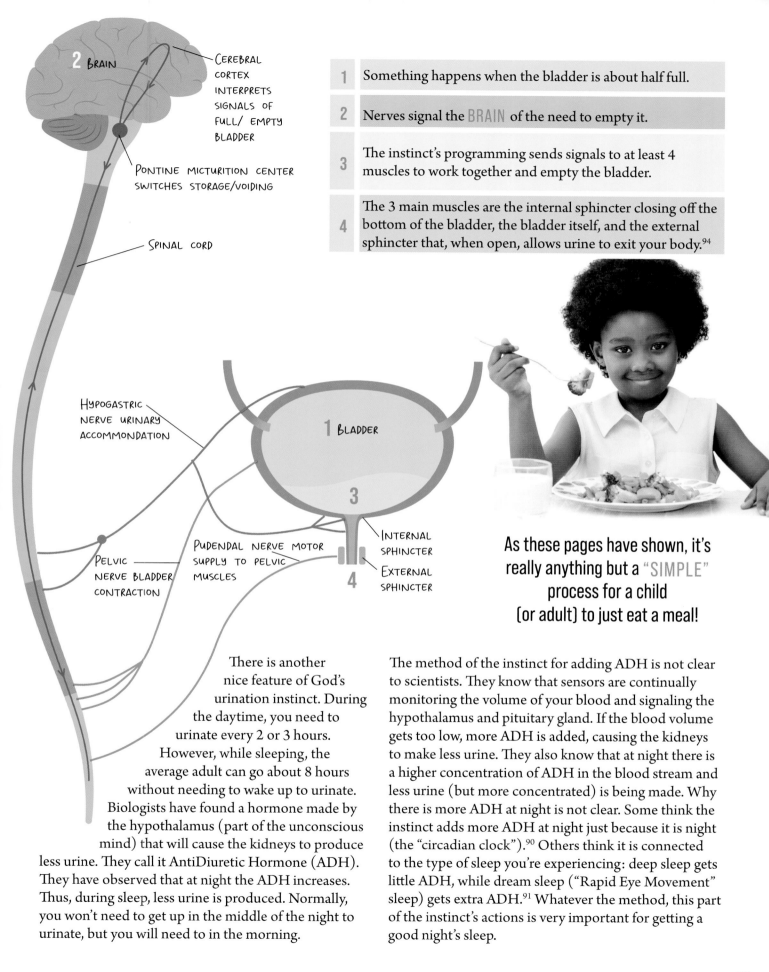

2 BRAIN

CEREBRAL CORTEX INTERPRETS SIGNALS OF FULL/EMPTY BLADDER

PONTINE MICTURITION CENTER SWITCHES STORAGE/VOIDING

SPINAL CORD

HYPOGASTRIC NERVE URINARY ACCOMMONDATION

PELVIC NERVE BLADDER CONTRACTION

PUDENDAL NERVE MOTOR SUPPLY TO PELVIC MUSCLES

1 BLADDER

3

4

INTERNAL SPHINCTER

EXTERNAL SPHINCTER

1	Something happens when the bladder is about half full.
2	Nerves signal the BRAIN of the need to empty it.
3	The instinct's programming sends signals to at least 4 muscles to work together and empty the bladder.
4	The 3 main muscles are the internal sphincter closing off the bottom of the bladder, the bladder itself, and the external sphincter that, when open, allows urine to exit your body.[94]

As these pages have shown, it's really anything but a "SIMPLE" process for a child (or adult) to just eat a meal!

There is another nice feature of God's urination instinct. During the daytime, you need to urinate every 2 or 3 hours. However, while sleeping, the average adult can go about 8 hours without needing to wake up to urinate. Biologists have found a hormone made by the hypothalamus (part of the unconscious mind) that will cause the kidneys to produce less urine. They call it AntiDiuretic Hormone (ADH). They have observed that at night the ADH increases. Thus, during sleep, less urine is produced. Normally, you won't need to get up in the middle of the night to urinate, but you will need to in the morning.

The method of the instinct for adding ADH is not clear to scientists. They know that sensors are continually monitoring the volume of your blood and signaling the hypothalamus and pituitary gland. If the blood volume gets too low, more ADH is added, causing the kidneys to make less urine. They also know that at night there is a higher concentration of ADH in the blood stream and less urine (but more concentrated) is being made. Why there is more ADH at night is not clear. Some think the instinct adds more ADH at night just because it is night (the "circadian clock").[90] Others think it is connected to the type of sleep you're experiencing: deep sleep gets little ADH, while dream sleep ("Rapid Eye Movement" sleep) gets extra ADH.[91] Whatever the method, this part of the instinct's actions is very important for getting a good night's sleep.

16

How quickly do instincts warn that something should not be swallowed?

Reject Bad Smell: Food that is good for a human will generally smell good to him or her. That is, as long as it really is good for him. As time goes by, if the food is not eaten by a human, it will probably be eaten by microbes (such as bacteria, yeast, and molds). As the microbes change the food into microbe waste (such as sulfur compounds), the food's smell and appearance will usually change.[95] "Foods affected this way will almost certainly not taste good."[96]

The human smell instinct will almost always make the human reject the food that has become bad for him. "People who have lost their sense of smell face a higher risk of food poisoning, often requiring hospitalization."[97] Think about how the human smell instinct can keep us from eating what would possibly kill us, while some animal's smell instinct makes it eat that same material... because it's good for that animal!

Reject Bad Taste: The amazing taste instinct, using the complex system of taste receptor genes, neurons, et al., both feeds us and protects us. "Thus, desired nutrients at appropriate levels can elicit pleasant tastes and harmful levels of toxins elicit very unpleasant tastes."[98] "If you ate poisonous or rotten foods, you would probably spit them out immediately, because they usually taste revolting."[99]

Your instinct will first determine if something should not be swallowed. It will then make sure it is not swallowed. This has to happen rapidly or it's useless. Instruments that scientists use to determine what an unknown material is (like a "food") almost never give any kind of answer within less than 5 minutes (usually a few hours, but sometimes many hours). Yet even babies' bodies can make the determination about thousands of different materials within 1 OR 2 SECONDS!

Remember how easy it is to swallow things that get into the mouth. Mothers are continually concerned about what babies put into their mouths. Yet, if some plant or piece of meat (normal kinds of food for humans) is really poisonous, as the scientist noted in the quote above, "you would probably spit them out immediately." Thank God for such a marvelous protection instinct.

Food Stuck in Intestine: It is fortunately rare, but food could get stuck in your intestine. However, it shows how well God made our bodies. Let's examine what happens if one of the intestines becomes blocked.

First realize how dangerous an intestinal blockage is. "A complete blockage is an emergency and needs immediate medical attention."[100] The reason is that pressure can build up as the peristaltic pumping continues pushing the stomach juices into the intestine. If the intestine ruptures, that acidic liquid full of waste material and bacteria would flood into the abdominal cavity and start damaging your sensitive organs. "This is a life-threatening complication."[101]

So, what do the instincts do to protect us? They use the signals from the many nerves in the affected regions to cause feelings of SEVERE pain, cramping, fullness, gassiness, etc.[102]

When the intestines are blocked, the body initiates various mechanisms to attempt to unblock them. These instinctual mechanisms can include	
1	muscular contractions (peristalsis) to push the blockage forward,
2	increased secretion of fluids to lubricate the intestines,
3	the activation of reflexes to stimulate bowel movements.

Amazingly, these same nerves are normally "silent" as they experience the almost constant motions of the stomach and intestines as they churn the food solution flowing through them. The instinct controls this. The pain will continue until the blockage is removed.

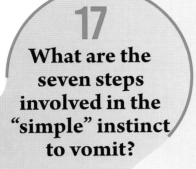

17

What are the seven steps involved in the "simple" instinct to vomit?

Another remarkable design in human bodies, even in babies, is the backup program (instinct) protecting us from bad food: vomiting.

God gave us the ability to judge food suitability by sight and smell. Even babies have the ability to taste whether a food is good to swallow.

However, sometimes, a food has spoiled and would not normally be swallowed, but our instincts are tricked. Salt, sugar, or other chemicals (such as MSG, which has the savory taste of protein while enhancing the other four tastes[103]) can hide that a food is bad. As a result, we may eat what is really not good for the body.

Fortunately, God created a complex system for emptying the stomach before much harm can be done to your body. This system, called vomiting, is very important, though you probably never liked using it. Look at this carefully designed method of cleaning out a human stomach. The following description is taken from the *BBC Science Focus Magazine*.[104]

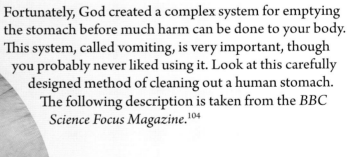

The vagus nerve and brainstem in the control of the vomit reflex.
1	First, signals are sent to the brainstem if any threats are detected. For example, the vagus nerve will detect an upset stomach. Other sensors can detect toxins or stress hormones in the blood. The inner ear can report too much swaying motion of the head (e.g., sea sickness). Once the brainstem's instinct determines that vomiting is needed, it sets off a "choreographed sequence of actions."
2	Second, the salivary glands start flooding the mouth with extra saliva, which will protect the mouth and teeth from the stomach acid that is about to come up and out. This extra saliva is also a warning to the conscious mind to prepare to vomit (i.e., quickly get to an appropriate location to empty your stomach).
3	Third, the human takes a deep breath, and the lungs are closed off by the epiglottis. This keeps the stomach contents out of the lungs.
4	Fourth, the sphincter muscle at the bottom of the stomach closes. This means the only way out of the stomach is up through the esophagus and throat.
5	Fifth, the diaphragm (the main muscle for breathing) works with the abdominal muscles "in a few short pulses, squeezing the stomach to create pressure."
6	Sixth, the stomach is squeezed and quickly emptied of its contents.
7	Seventh, the brainstem speeds up the heart rate to make the whole body sweat. Why? Scientists have determined that this is to "shed the heat from this sudden exertion." God thought of everything.

You probably thought vomiting was a simple action.

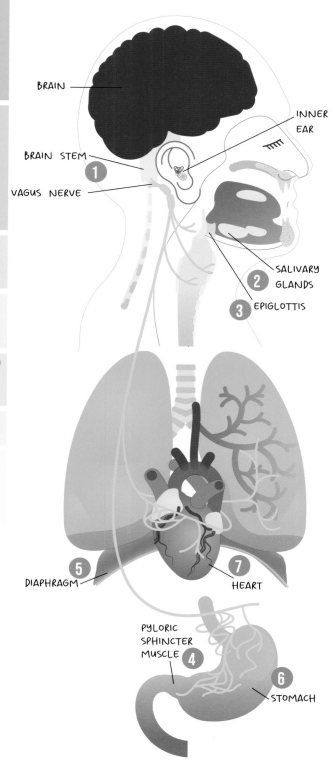

THE VAGUS NERVE IS RESPONSIBLE FOR THE REGULATION OF INTERNAL ORGAN FUNCTIONS, SUCH AS DIGESTION, HEART RATE, AND RESPIRATORY RATE, AS WELL AS VASOMOTOR ACTIVITY, AND CERTAIN REFLEX ACTIONS, SUCH AS COUGHING, SNEEZING, SWALLOWING, AND VOMITING.

BRAIN

INNER EAR

BRAIN STEM

VAGUS NERVE

SALIVARY GLANDS

EPIGLOTTIS

DIAPHRAGM

HEART

PYLORIC SPHINCTER MUSCLE

STOMACH

18

Why is the instinct to sleep one that we cannot simply override?

Most of us probably view sleep as a waste of time that could be used for either fun or work. However, scientists have confirmed that sleep is vital to our health. That is why God built into humans the sleep instinct. Apparently, it is continually monitoring our bodies to determine when we need to do more than just "rest," and when it is okay to stop sleeping (wake up).

How important is sleep? This detailed scientific article describes a multiplex process (using at least 6 structures within the brain) that scientists have concluded is essential for life! Key points are bolded in the following quote from the article:

"**Quality sleep** — and getting enough of it at the right times — **is as essential to survival as food and water.** Without sleep you **can't** form or maintain the pathways in your brain that let you learn and create new memories, and it is **harder to concentrate** and respond quickly.

Sleep is **important to a number of brain functions,** including how nerve cells (neurons) communicate with each other. In fact, your brain and body stay remarkably active while you sleep. Recent findings suggest that sleep plays a housekeeping role that **removes toxins** in your brain that build up while you are awake.

Everyone needs sleep, **but its biological purpose remains a mystery.** Sleep **affects almost every type of tissue and system in the body** — from the brain, heart, and lungs to metabolism, immune function, mood, and disease resistance. Research shows that a chronic **lack of sleep,** or getting poor quality sleep, **increases the risk of disorders** including high blood pressure, cardiovascular disease, diabetes, depression, and obesity.

Sleep is a **complex and dynamic process that affects how you function** in ways scientists are now beginning to understand. This booklet describes how **your need for sleep is regulated** and what happens in the brain during sleep[105] (emphasis added).

A newborn baby's body has the sleep instinct that forces it to sleep. Adults who stay up too many hours in a row will eventually fall asleep, no matter how much they resist this essential instinct! God obviously cares for us since He created sleep as an instinct that we cannot completely override.

? Sleep seems like a waste of time, so are scientists working on helping us go without sleep?

"The National Sleep Foundation's recommendations for nightly sleep are broken down into nine age groups (see chart below) … How much sleep *you* need means considering your overall health, daily activities, and typical sleep patterns."[106] Think about that statement. That means this sleep instinct is so sophisticated that it adjusts for each individual at the age he happens to be at the time! This complex instinct does not act as if all people need the same amount of sleep.

	Age Range	Recommended Hours of Sleep
Newborn	0–3 months old	14–17 hours
Infant	4–11 months old	12–15 hours
Toddler	1–2 years old	11–14 hours
Preschool	3–5 years old	10–13 hours
School-age	6–13 years old	9–11 hours
Teen	14–17 years old	8–10 hours
Young Adult	18–25 years old	7–9 hours
Adult	26–64 years old	7–9 hours
Older Adult	65 or more years old	7–8 hours

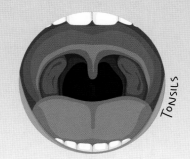
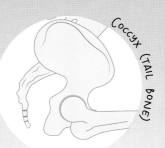

APPENDIX · TONSILS · COCCYX (TAIL BONE)

True Science Knows There Was a Creator

Please observe that scientists know that EVERYTHING has a purpose. In the long quote on the previous page, they say, "but its biological purpose remains a mystery." They know there is a purpose, so they are still looking for it.

The only plausible theories of origins are an outside Creator (the God of the Bible) and evolution (no outside Intelligence). If there was a Creator, a scientist should expect a purpose for everything. If there was no Creator (evolution), why would a scientist expect a purpose?

Evolution teaches that there should be many "vestigial organs" in all creatures. A vestigial organ would be one that originally had a purpose, but the evolving of the species caused it to not be needed anymore. Therefore, evolution teaches that many parts of creatures should have no purpose now. Around 1900, some evolutionists said there were 122 vestigial organs in a human body. Real science has shown that almost all of those are useful (such as the appendix, the tonsils, and the coccyx). Today, most evolutionists will only claim up to 6 vestigial organs, but even all of those are being debated. So, even the evolutionists are being forced to think everything really might have a purpose now.

Scientists know from their experience that there is ALWAYS a purpose!

What might be the most important instinctive reflex a baby has?

Temporary Instincts for Human Babies

In God's design are some programs (instincts) that disappear within the first year of life. Some seem to be for protection of the almost defenseless child, and others seem to assist the child in learning skills that will be used the rest of his life. Let's look at some of them.

Grasping Reflex: One instinct for protection would be the grasping reflex where the child will wrap his fingers around anything that touches his palm. From the minute he exits the womb, his fingers will close on and grasp almost anything. Some researchers think this instinct is important because "baby needs mom close to grow and feel secure."[107] Other scientists think it "helps babies develop the skills to intentionally grasp things as they grow."[108] Whatever the reasons, they know this programming is not an accident; it is a design.

It disappears within 12 months.

Rooting Reflex: A newborn baby will instinctively turn his head toward sounds, smells, touch, or visual cues. Why? He is looking for his mother's milk.[109] God has given newborns a strong sense of smell and they know the scent of their mother's milk. He will keep moving his head until he finds the nipple.

Notice that the nipple is not what the baby actually needs. He needs the milk. So, this instinct knows that a nipple, though it may smell like milk, is not milk; it must be sucked on. That is as amazing as a Viking from a thousand years ago (before food canning was invented) being teleported into a grocery store, seeing a picture of a fish on a sardine can, and knowing he must lift the ring up and pull the lid off the can of sardines to get the fish.

One researcher called this instinct a "method of survival."[110] In other words, without this programming, a baby's life would be in immediate danger.

Within the first 4 months of life this reflex disappears as the conscious brain takes control of our desire to eat.

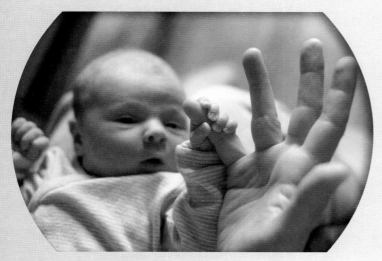

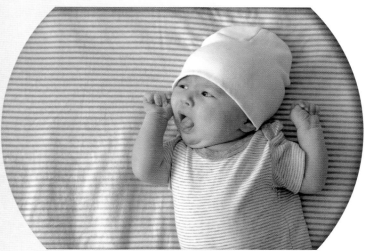

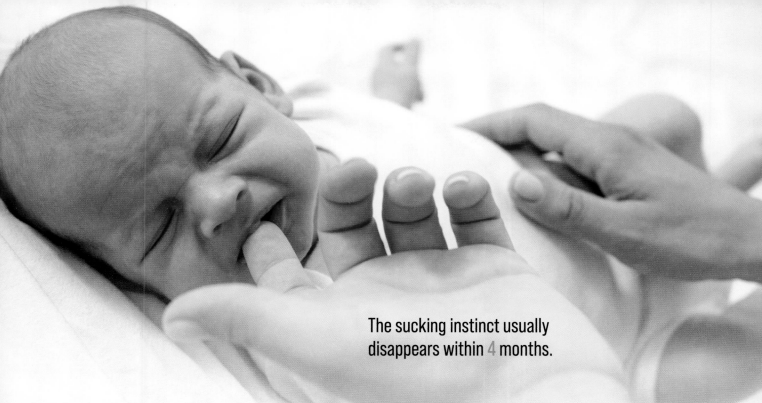

The sucking instinct usually disappears within 4 months.

Sucking Reflex: The ability of a newborn to get food into his body is amazing. If he doesn't succeed, then he will die.

God not only designed this ability to suck milk out of his mother's breast, He started its operation around 8 weeks before the normal time to exit the womb (32 weeks of gestation).[111]

Sonograms of a baby in the womb will often show him sucking his thumb. Why would God start it so early? Probably because it's not uncommon for babies to not stay in the womb for the full 40 weeks. If a baby comes out after only 35 weeks, how would the mother feed him? If God didn't start the sucking instinct at about 32 weeks, those premature babies would most likely die of starvation as the mothers desperately tried to figure out how to feed them.

The sucking reflex "might be **the most important reflex**, since it helps baby suck and swallow milk…. Interestingly, this reflex is one of the only behaviors that **is more skilled in babies than it is in adults"**[112] (emphasis added). This fact is another strong statement that the Designer knows our needs and provides for all of them.

Notice how an evolutionist sees this need and the design that meets the need, and it leads him to conclude that since there's no God and the instinct exists, then SOMEHOW, SOMETHING made it: "It is easy to see how

this behavior evolved. It increases the chances of a baby feeding and surviving."[113] The idea that babies would never survive without the sucking reflex does not seem to shake the FAITH of the evolutionist who believes that they somehow DID survive during a hypothetical time when the reflex didn't yet exist.

The rooting reflex mentioned previously will cause the baby to find the mother's breast and nipple. The sucking reflex then gets the milk out and into the baby's stomach. Both are important.

The sucking reflex consists of 3 stages.	
1	First, he places his lips over the nipple and onto the breast. His lips are wet with saliva and make an air-tight seal. He then begins the process of sucking by pulling a vacuum with his lungs. As he does this, he squeezes on the breast with his tongue.
2	Second, he quickly moves his tongue to press on the nipple. This pressure and the suction of his lungs pulls the milk out and into his mouth.
3	Third, the baby swallows the milk.

"**Coordinating these rhythmic sucking movements with breathing and swallowing is a relatively complicated task for a newborn**"[114] (emphasis added). Newborn babies are unable to do much of anything, yet they can do this very well.

20

What is the Landau Reflex that helps babies develop correct posture?

I will praise thee; for I am FEARFULLY AND WONDERFULLY MADE: *Marvellous are thy works; And that my soul knoweth right well. (Psalm 139:14)*

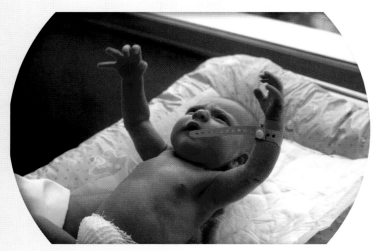

Tongue-Thrusting Reflex: There's another important reflex that you've probably never noticed: tongue-thrusting. It is in all, or almost all, newborns. It is seen when anything other than a nipple touches their tongue. The reflex makes the tongue push any object out of the mouth … unless it is a nipple!

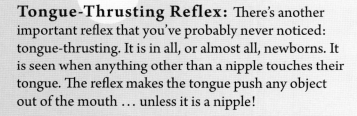

	This simple behavior of Tongue-Thrusting has several important purposes that scientists have discovered.
1	One is that it "helps protect babies from choking or aspirating food and other foreign objects."
2	Another is that it "helps them to latch onto a nipple." This prevention of anything but milk getting into his mouth is important "because their muscles aren't yet developed enough to swallow anything more than liquid."
3	Lastly, this instinct starts to disappear at about 5 months, which is when a baby's body is ready to start eating solid foods.[115]

It is a simple, but important design, disappearing at about 5 months.

The Startle Reflex: A sensation of falling, "loud noises, intense light, and sudden movements can trigger a baby's" startle (or Moro) reflex.[116] The baby will throw his arms out straight and usually throw the head back and extend the legs. Once the baby feels safe, when the "trigger" has stopped and he feels secure, he will curl the arms and legs back toward the body and relax.[117]

If the baby is falling, after the initial throwing out of the arms and legs, his "**instincts take over** and assume the fetal position — in order to best brace for impact of a fall. So amazing when you think **of how a baby is wired** to try to protect himself during falls"[118] (emphasis added).

As always, the researchers know there is a purpose for everything God created. They say this reflex is "meant to protect baby from danger. Baby is learning to respond to something that seems dangerous" and to use his body to protect himself.[119]

Once the baby's neck can support the head, at about 4 months of age, he starts having fewer and less intense startle reflexes.

They are usually completely gone by the age of 6 months.[120]

Muscle development to learn to walk, eat, and crawl starts long before the baby is able to do so.

Stepping Reflex: You've probably seen a parent lift a 2-month-old baby to an upright position, let his feet touch the floor, and watch his legs move as if he were walking. What good is it, since it goes away at about 3 months, and real walking won't occur for almost a year? You may rightly wonder why God created this instinct.

Since even evolutionists know there is a purpose for everything, they have studied it to determine why it exists. "This reflex will help a newborn crawl to the breast right after delivery when lying on their mother's abdomen."[121] "Ten-week-old babies can learn from practicing walking months before they begin walking themselves, say researchers."[122]

"Experts say this reflex ensures the leg muscles develop properly, so baby can learn to walk."[123]

Goes away at about 3 months.

Learning to Feed Himself: As the baby gets older, the hand-to-mouth reflex (instinct) becomes important. It is apparently designed to prepare the baby for feeding himself. It also gives the baby a sense of calm, gives him awareness of all his fingers, and builds hand-eye coordination. With use, this hand-to-mouth reflex leads to the baby's brain developing the ability to easily perform self-feeding.[124]

Reflex for Correct Posture: A rather strange reflex is called the Landau Reflex. It starts at about 3 months of age and continues until the baby is about one year old. It is seen when the baby is in a horizontal position, face down, either suspended by adult hands or lying on a flat surface. The baby will raise his head, flex his back, and start flexing his legs. Before a baby is able to crawl, he can often be seen on the floor doing these motions.

Experts say this reflex "helps baby stand up straight and develop correct posture."[125] The reasoning is that he is working the muscles in his neck, back, and core.

Goes away at about 12 months.

21 How do babies acquire so many skills so early on?

Learning to Walk: The next stage after learning to stand is to learn to walk. The parents will get on the other side of the room and encourage the baby to walk to them. That is all the parents do. The walking instinct takes control, coordinating the weak muscles to attempt to carry the baby's weight across the room.

Walking would be much simpler if humans had wheels instead of legs. However, we would be much more limited in where we could go. Since we have legs with feet on the bottoms, walking is complicated. The instincts use at least 8 sets of muscles to walk without falling on our face.[126] Timing must be correct to keep moving while maintaining balance.

Walking is so complicated that scientists still do not fully understand how humans, including children, can do it so well. "It's never been completely clear how human beings accomplish the routine, taken-for-granted MIRACLE we call walking, let alone running.... Human walking is **extraordinarily complex** and we still don't understand completely how it works..."[127] (emphasis added).

? Why do small children learn a second language so easily compared to an adult?

With the start provided by the walking instincts, the brain's neurons are connected in the ways necessary to refine our walking ability and make it able to be done with little mental concentration. The next page explains this in more detail.

A general overview of the key muscle groups involved in maintaining balance and stability during walking	
1	Leg Muscles
2	Gluteal Muscles
3	Abdominal Muscles
4	Back Muscles
5	Hip Muscles
6	Foot and Ankle Muscles
7	Shoulder Muscles
8	Arm Muscles

Instinct for Learning New Skills: "Babies enter the world with a grasp of some pretty complex topics — and how they build on that and what they continue to learn in the first two years is totally amazing too. Just a few years ago, most experts and parents believed the way tots learned was random and passive — a slow process of pint-sized trial-and-error. Turns out **babies are born learners who come into this world with a natural understanding of all sorts of things,** including ethics, physics, and language"[128] (emphasis added). This quote says nicely one of the realities this book is trying to show: God has created human babies with the built-in tools (instincts) necessary to learn the things needed to live a healthy life.

Studies have shown how God has made human children programmed for learning. A 2019 report states that, "the **brain consumes** a lifetime peak of two-thirds of the body's resting energy expenditure, and **almost half of total expenditure,** when kids are **five years old** … (and) the **energy needed for brain development declines in older children** and adolescents…."[129] (emphasis added). The adult brain's usage is only 20% of the body's resting energy usage.[130]

Researchers concluded that the developing brain was like a computer that was being built (hard-wired) to do certain activities. After age 11, it is more difficult for the brain to learn new activities, because the "hard-wiring" is finished and only "software" can accomplish a new task. If you know anything about computers, you know very little software is needed if a computer is hard-wired to do a specific task. That explains why little children can easily learn a new language compared with adults (who often just give up because it is so hard).

Newer studies have shown that the "hard-wiring" of the brain is the connecting of the neurons to each other. A newborn baby's brain has all of the neurons it will ever have. However, there are almost no connections between the neurons. One study found, "The early childhood years are crucial for making these connections. At least one million new neural connections (synapses) are made every second, more than at any other time in life…. It's much harder for these essential brain connections to be formed later in life."[131]

Do you see how amazing this process is? The baby does not decide he needs to spend more energy in building his brain! What a marvelous design that takes the energy and develops a baby's brain to easily do so many different activities!

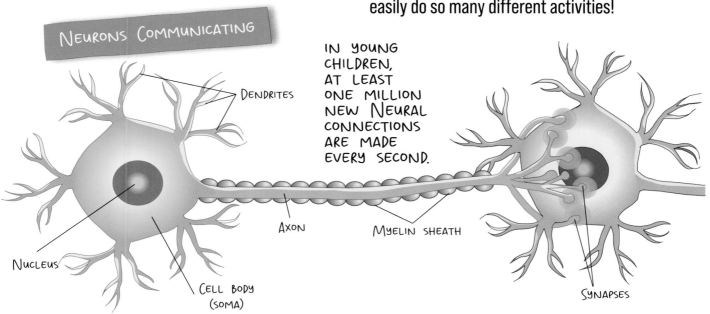

NEURONS COMMUNICATING

IN YOUNG CHILDREN, AT LEAST ONE MILLION NEW NEURAL CONNECTIONS ARE MADE EVERY SECOND.

DENDRITES

AXON

MYELIN SHEATH

NUCLEUS

CELL BODY (SOMA)

SYNAPSES

A child is born with about 100 billion brain cells (neurons). At birth, the average baby's brain is about a quarter of the size of the average adult brain. Incredibly, it doubles in size in the first year and keeps growing to about 80% of adult size by age 3 and 90%, nearly full grown, by age 5.

22

What can a blue dot teach us about the amazing design of the eye?

Overview of the eye: The human eye has always bothered people who believe in evolution.[132] It certainly looks designed by an intelligence, rather than by dumb luck as evolution claims. This book will concentrate on the programming needed to make your eyes work.

We have a little control over the focusing of our eyes, but not much. Most of the focusing of our sight is done by part of our subconscious mind (the part of the brain below the cerebrum, called the pons).[133]

If you think about it, it is really wonderful how we can rapidly look at many different objects at many different distances away and still very quickly see each in focus. Life would be difficult without this programming (instinct) for focusing our vision.

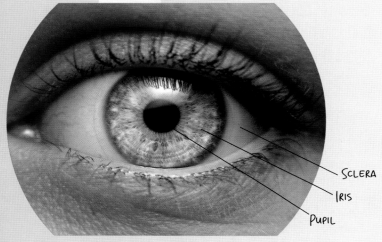

A very simplified description of the eye and how it focuses on objects would be as follows.

Consider a tiny, blue dot on the page of this book. Light from that blue dot flies out in every direction. The whole, exposed surface of your eye is hit by that blue light. The blue light rays that hit the eye's curved, transparent cornea will go through but be bent toward each other because of the curvature. These rays then enter and go through the iris and then the lens of the eye. This lens is curved just the right amount by your instincts to cause all of those blue light rays to meet at the back of the eye to make one tiny blue dot. The back of the eye, the retina, is covered by cells that sense the

blue dot and report it to the brain as a picture of that tiny, blue dot on this page.

Now let's examine just some of the amazing aspects of how even a child is able to "see" that blue dot.

The Iris: First, it is possible to overwhelm the retina's sensory cells, the rods and cones, with too many light rays. Therefore, God has put between the cornea and the eye's lens an adjustable hole the light must go through. This hole is called the pupil. This hole is created by the iris, which is a circular screen with a circular hole in the middle of it. Two muscles move the iris so that the pupil is always in the center of it and always a perfect circle. If the hole could move from side to side, vision would be much harder to control.

When the intensity of light changes, the instinct is able to rapidly move the iris to allow the correct amount of light onto the retina for the rods and cones to send a clear picture to the brain. The instinct uses the information from the rods and cones to adjust the size of the pupil. The instinct quickly calculates the size needed and commands the two muscles to move the iris.

By now, you should not be surprised that the details of how the instinct adjusts your pupil size are more complicated than the above description. There are several chemicals that are used for controlling the two muscles. Also, the instinct will make its calculations based on both eyes and always make your two pupils the same size.[134]

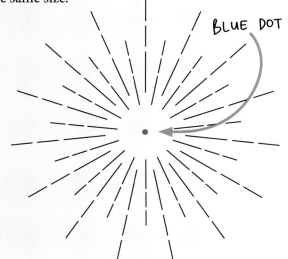

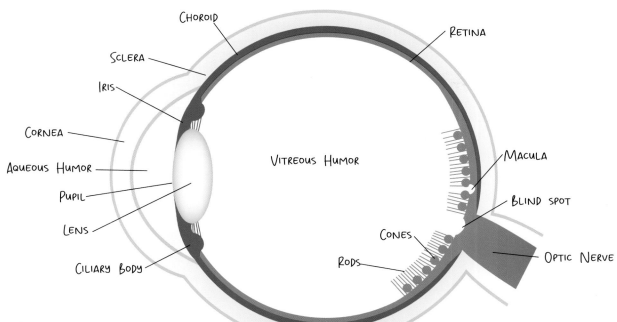

The Lens: After the light rays go through the pupil, they enter and pass through the lens. The lens is a very complex set of transparent cells. They are individually shaped and stuck together to make a soft, clear ball that can be pulled and stretched for many years without the cells ever separating from each other. They also go for many years without losing their transparency.

Unlike almost all other cells in your body, these cells do not have nuclei. The lens would be less transparent if its cells had nuclei. Evolutionists must BELIEVE the nuclei in ALL of these cells would be removed without a Designer!

The changing of the curvature of the lens is marvelously done by your instincts. As the ciliary muscle relaxes, the zonular fibers pull on the lens and it becomes thinner, the closer objects become blurry and those farther away become clearer ("in focus"). As the ciliary muscle contracts, the zonular fibers relax and the lens becomes more round, the closer objects come into focus.

The design of the cells of the lens and the muscles that pull on them is such that the changing curvature will always cause a shape that amazingly will produce an undistorted picture on the back of the eyeball of objects a similar distance in front of the eye. In other words, no matter how much the lens shape is changed, if you cut the lens in half, horizontally, right through its middle, the two halves' curvatures, up and down from the middle, would be exact duplicates of each other. If this weren't true, your vision could easily become like images produced by the warped mirrors at the House of Mirrors in an amusement park.

Your conscious brain does not tell the muscles how much to pull on the lens. This is done by the programming of your instinct.

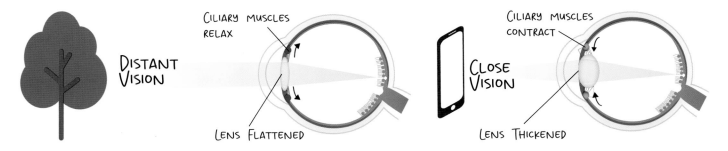

CILIARY MUSCLES RELAX AND ZONULAR FIBERS PULL THE LENS AND FLATTEN IT TO FOCUS ON DISTANT OBJECTS.

CILIARY MUSCLES CONTRACT AND ZONULAR FIBERS RELAX AND ALLOW THE LENS TO THICKEN TO FOCUS ON A CLOSE OBJECT.

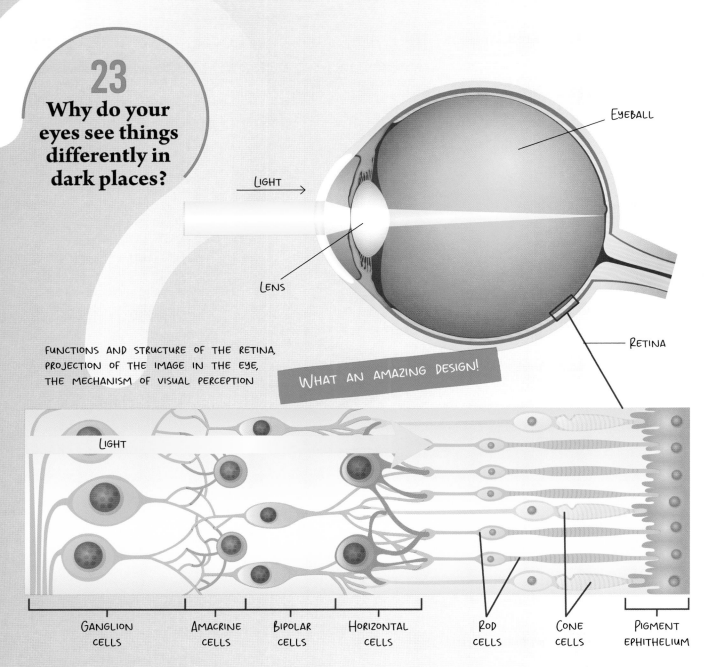

23
Why do your eyes see things differently in dark places?

LIGHT →

EYEBALL

LENS

RETINA

FUNCTIONS AND STRUCTURE OF THE RETINA,
PROJECTION OF THE IMAGE IN THE EYE,
THE MECHANISM OF VISUAL PERCEPTION

WHAT AN AMAZING DESIGN!

LIGHT

| GANGLION CELLS | AMACRINE CELLS | BIPOLAR CELLS | HORIZONTAL CELLS | ROD CELLS | CONE CELLS | PIGMENT EPHITHELIUM |

Retina: The lens projects an image onto the back of the eyeball. The back of the inside of the eyeball is covered by several layers of important materials. The layer directly related to vision is called the retina. It is full of an enormous number of light sensitive cells. One human eye has about 90 to 120 million rods and about 7 million cones.[135] Each light-sensitive cell is remarkable, and, as you should expect, they are arranged in the eye to give us very good vision.

The eyeball is designed so that a human can see a complete picture of almost a full half sphere in front of him, and it's all in color. This is because the light-sensitive retina covers almost a full half of the back of the eye. However, the half-spherical image is clearest and most colorful at its very center (directly in front of each eye). The farther away from the center, the less clear and less colorful the image is. This variation is due to the arrangement of the light-sensitive cones and rods.

The cones detect all the colors that humans can see. Your cones determine your ability to enjoy a beautiful painting. They also provide your ability to read the fine print and see far away. This is because each cone will report separately to the brain what it sees. The other light sensitive cells, rods, report to the brain in groups.[136] Thus each cone is like a pixel on your television screen or computer monitor, while a group of rods become like just one pixel.

The specialty of the rods is their sensitivity to even one photon of light.[137] The rods are much more sensitive to light than cones are (one scientist says 100 times more sensitive,[138] while another says 1,000 times[139]). The rods don't report color to the brain, only light intensity, which is interpreted as shades of gray. So the rods allow you to see in very dark places; but it will be like a black and white picture!

As usual, it is more complicated than what has been described. The arrangement of the cones and rods is remarkable. GOD HAS DESIGNED THE HUMAN EYE TO HAVE THE HIGHEST CONCENTRATION OF CONES (NO RODS AT ALL) IN THE VERY CENTER OF THE IMAGE THAT THE RETINA DETECTS. Thus, when you try to "focus on something," you will stare at the object whose image is hitting the fovea centralis (the middle of the retina in the back of the eye where the most cones are). Those cones will give the best resolution (number of pixels) and the best color image.

This circular area at the center of the retina with many cones and no rods (the fovea centralis) is only 0.3mm in diameter.[140]

As you get away from that "focal point," the number of cones rapidly decreases as the number of rods rapidly increases. The total number of light-sensitive cells stays high from the center to about halfway to the edge of your whole, half-spherical image. The total number keeps decreasing but is still high at the edge of your half-spherical image.

This explains several things. IN A VERY DARK PLACE, YOU CAN STILL SEE IMAGES, BUT NO COLOR. THAT'S BECAUSE THERE'S NOT ENOUGH LIGHT FOR THE CONES TO REACT AND GIVE YOU COLORS. THE GROUPS OF RODS ARE REACTING TO GIVE A BLACK AND WHITE IMAGE WITH POOR RESOLUTION. Also, since there are no rods in the focal point, in the dark you can't see a small object unless you move your eye to get the object out of that focal point and onto some rods—then you can see it! This is well-known by astronomers trying to see very faint stars.[141]

Even though the eye is marvelously designed to do amazing things, it still needs some more special programming (instincts) for its routine operation. Each eyeball's retina has a "blind spot" where its optic nerve connects to the retina. You have probably closed one eye, looked at a very small object, and slowly moved your eye toward your nose; the small object would have disappeared when its image fell onto the optic nerve ending. Your instincts will continually "imagine" what the missing images are that are landing on the optic nerve ending (the "blind spot"). If, however, the missing image is not an extension of a surrounding larger object or pattern, your brain is unable to create a correct image … and the small object will "disappear." This instinct is so good that you probably never knew you had a "blind spot" until someone showed you.

Other instincts are also continually working to allow you to see things. The instincts built into your brain know how to adjust the iris for the right amount of light going through the pupil. You know how quickly your iris adjusts when the light intensity changes; otherwise, this momentary "blindness," if longer, could become dangerous. At the same time, the instincts are pulling on the lens to adjust the focal distance to be where the object of interest is. When the object of interest is at the focal distance, the cones in the fovea centralis will present to the brain a focused picture of it (the brain will "see" it clearly).

? If you want to see a distant object in the dark, you won't see it if you look directly at it; you must shift your focus just to the side of it. Why?

LESS COLOR
WALKING IN THE DARK

24

How can our eyes create 3-dimensional versions of objects?

Using Both Eyes for Depth Perception and 3-D Imaging: Individual eyes can provide the brain with a 2-dimensional picture. However, using them together, your instincts can determine the distance to an object and even create a 3-dimensional version of the object.

Even babies need to know the distance to objects around them. This is mainly done with the eyes working together. You have probably experimented with trying to tell the distance to something with just one eye open and laughed at your failure. It is not a laughing matter to be unable to know how far things are away from you. You use that ability all the time. Have you ever considered how difficult it really is to calculate the distance to an object?

The primary method of determining distance to an object is using parallax. The brain constructs a triangle (left eye — right eye — object) and calculates the distance to the object. Again, it is amazing that the brain knows the precise distance between the pupils of the two eyes; without this distance, the calculation would fail. The other two measurements necessary for the calculation is the exact direction each eye is looking. How can the brainstem know those directions so precisely? Yet even a baby's brainstem can do it.

The instinct God put into humans for determining distance is more complex than just using parallax. It uses at least two other pieces of information. First, if you have an idea of how large the object actually is, the brain will use that to refine the result of parallax. Second, the brain can have you move your head back and forth to effectively increase the distance between your eyes; thus, the base of that parallax triangle becomes larger, and the precision of the parallax triangle's angles becomes greater.[142]

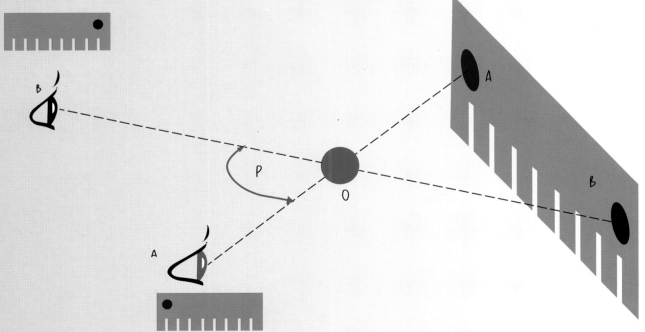

THE PARALLAX EFFECT CAN BE DEMONSTRATED WITH A SIMPLE ILLUSTRATION. IMAGINE AN OBJECT PLACED AGAINST A DISTANT BACKGROUND. WHEN OBSERVED FROM "VIEWPOINT A," THE OBJECT APPEARS TO BE POSITIONED IN FRONT OF A BLUE CIRCLE. HOWEVER, WHEN THE VIEWPOINT IS SHIFTED TO "VIEWPOINT B," THE OBJECT SEEMS TO HAVE MOVED AND NOW APPEARS IN FRONT OF A BLACK CIRCLE. THIS CHANGE IN PERSPECTIVE CREATES THE ILLUSION OF THE OBJECT'S POSITION SHIFTING RELATIVE TO THE BACKGROUND ELEMENTS.

Because human eyes are relatively close to each other, the parallax calculation of the brain begins to lose its effectiveness when an object is more than 18 feet away. Beyond that distance the brain begins to depend more and more on the other distance-determining methods: prior knowledge of an object's size and moving the head back and forth.[143]

Besides calculating the distance to an object, a correctly functioning human brain will create a single 3-dimensional (3-D) image out of the two 2-dimensional (flat picture) images from your two eyes. "The mind combines the two images by matching up the similarities and adding in the small differences."[144] Your brain's 3-D image will allow you to "see" depth in the object itself; some parts will "look" closer than other parts. It is all the result of another complicated math calculation that God has put into the brains of humans, even babies.

Why does looking across the approximately 10-mile-wide Grand Canyon give you the impression of a 2-dimensional picture postcard? The distance is too great for parallax to work. The distance is so great that moving your head back and forth does nothing to improve the parallax calculation. Almost nothing you can see on the far side of the Grand Canyon has a size known to your brain. Finally, the two images from your two eyes have no differences between them for the 3-dimensional calculation. The result is a surprising 2-dimensional image of the obviously 3-dimensional Grand Canyon!

This marvelous, instinctive use of both eyes to determine distance and create 3-D images in almost every situation can be done very well by little children. Yet, listen to what the researchers say. "They help us to see in three dimensions, recognize depth and use that information to calculate distances. **Exactly how the brain does this**, dividing space into categories such as 'within reach,' 'near,' or 'far,' is **largely unknown**"[145] (emphasis added).

? If you've stood on the edge of the Grand Canyon, why does the other side look flat like a postcard rather than look 3-dimensional?

25

Why are your ears not simply nice, smooth cups?

Instincts can use your ears just like your eyes are used to determine the distance to and direction of a sound source. However, it is even more complicated with the ears than with the eyes.

We all have ears, but have you ever thought about how difficult it would be to calculate the direction and distance to a sound source? It is unbelievably difficult, yet even children can do it, due to God creating the marvelous instinct that does it for them. The following is a summary of pages of details in three reports about directional hearing.[146, 147, 148]

Just to get the horizontal direction is incredibly difficult. The direction cannot be determined with only one ear. Two ears, separated by a known distance, must hear the same sound from the unknown direction. The precise time the sound wave hits each ear must be measured. Then the brainstem (unconscious part of the brain) will calculate the approximate shape of the triangle (left ear—right ear—sound source) and thus know the horizontal direction to the sound source (since your ears are connected by an imaginary horizontal line of that triangle with the sound source). Like with the eyes, this is parallax.

If a sound source is on one side of the baby, the sound wave will hit one ear before the other. Besides somehow knowing the exact distance between the ears, the brainstem of a human is able to measure the difference in arrival time of the same sound wave to each ear even if only 10 micro-seconds apart! Amazingly, the brainstem very quickly calculates the shape of that triangle and determines the direction to the sound source. This is done in even very young children!

Directional hearing is more astonishing than what has already been described. When you throw a rock into a calm lake, waves will emanate from it as growing circles. If you drop the rock into the water close to objects on the surface of the water, the waves will reflect off the objects without slowing down. You can create a very confusing pattern of waves on the surface of water by having several objects in the way of the growing circles.

The confusing pattern of water waves is much closer to reality for sound waves in air. They are continually bouncing off objects into your ears. They even bounce off various parts of your own body into your ears. Your brainstem can separate those extra signals from the key ones coming directly from the sound source!

Again, it is even more amazing than what has already been said. Have you ever wondered why your ears are not like you would design them: nice, smooth cups? God purposely made the irregular shapes of our ears to provide known places for sound waves to bounce off of and into the ear. The brainstem knows exactly where those irregular shapes are and can use those signals to fine tune its calculations of direction to the sound source!

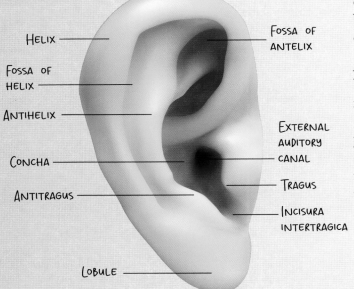

Helix
Fossa of Helix
Antihelix
Concha
Antitragus
Lobule
Fossa of Antelix
External Auditory Canal
Tragus
Incisura Intertragica

The outer ear, with its pinna and ear canal, helps collect and modify sound waves, while the inner ear, with the cochlea, translates these vibrations into electrical signals. The brain then processes the information from both ears to determine the direction of the sound source, enabling us to instinctively locate and perceive sounds in our environment.

BRAINSTEM (BREYN-STEM) NOUN: BACK SIDE OF BRAIN, CONNECTING THE CEREBRUM WITH THE SPINAL CORD. THIS IS MADE UP OF THE MIDBRAIN, THE PONS, AND THE MEDULLA OBLONGATA.

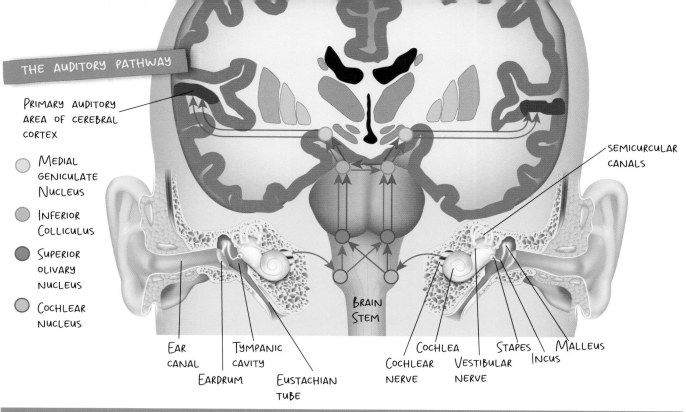

PRIMARY AUDITORY AREA OF CEREBRAL CORTEX

SEMICIRCULAR CANALS

○ MEDIAL GENICULATE NUCLEUS

○ INFERIOR COLLICULUS

○ SUPERIOR OLIVARY NUCLEUS

○ COCHLEAR NUCLEUS

BRAIN STEM

EAR CANAL

TYMPANIC CAVITY

EARDRUM

EUSTACHIAN TUBE

COCHLEA

COCHLEAR NERVE

STAPES

VESTIBULAR NERVE

INCUS

MALLEUS

A guide to how your ears work

STEP 1	STEP 2	STEP 3	STEP 4
SOUND TRAVELS INTO THE EAR CANAL UNTIL IT REACHES THE EARDRUM	THE EARDRUM VIBRATES AND SENDS SOUND THROUGH THE MIDDLE EAR BONES (OSSICLES) AND INTO THE INNER EAR (COCHLEA).	THOUSANDS OF TINY HAIR CELLS IN THE INNER EAR CONVERT THE SOUND INTO ELECTRICAL SIGNALS THAT ARE SENT TO YOUR BRAIN.	THE BRAIN PROCESSES THOSE SIGNALS AND TELLS YOU THAT YOU ARE HEARING A SOUND AND WHAT THAT SOUND IS.

Human directional hearing is still more intricate. The brainstem very quickly takes into account many signals to determine direction and even distance to a sound source. Things such as volume (intensity) in each ear, position of the head, previous knowledge of similar sound sources, visual cues, whether you cupped your hands over your ears, etc.

It even takes into account how much the human's head acts as a shadow reducing the intensity of a sound wave reaching the hidden ear. Even this is compounded by the fact (known to God's instinct's calculation) that the higher the sound frequency the more the head will hide it.

HAIR CELLS IN THE INNER EAR

If you care to see the complexity of the human directional hearing instinct, you can find a detailed presentation in Wikipedia under "Sound Localization" (typing "Directional Hearing" will redirect you to it).

The baby's brain uses his directional hearing to easily locate the person trying to communicate with him. Directional hearing is important enough to babies that when scientists tricked them by making the mothers' voices come from a different direction from where the babies' eyes saw them, the babies got "very upset." So, this complex instinct is very crucial for the development of the baby.[149]

26
Why would oxytocin release when a baby and mother make eye contact?

Another important instinct God has put into babies is the desire to communicate. Since the two greatest commandments from God, according to Jesus in Matthew 22:36-40, are to love God and love our neighbor, it is easy to see why God gave humans the instinct for communication. Communication helps in relating to God and to our fellow humans ("neighbors").

Designed to Hear Other Humans: Research has shown that newborn babies are designed to hear other humans. This was studied in several different ways.

Auditory-visual coordination was shown by a test using a mother's image and voice. The baby's mother was able to be seen through a sound-proof window, while the mother's voice was heard through several sets of speakers. When the voice came from the same direction as the mother's image, the baby was content. However, as mentioned previously, moving the voice elsewhere, especially from behind, the baby became very upset. This directional ability of even newborn babies is important for human interaction.

? What happens when babies and their mothers stare at each other's eyes?

Another study showed that even in the first days of life outside the womb, babies can distinguish between human and computer-simulated crying. The researchers watched as newborns listened to human baby crying and became much more restless than when hearing the fake crying.

Other studies showed a preference in newborns for the human voice over a nonhuman voice. This is also important. The instincts God put into humans encourage us to listen to each other.

A different study showed that newborns prefer to hear their own mother over any other voice. This is of course necessary for a child to be trained and protected by his own mother, whose instincts guide her to want to train and protect him.

These studies show the built-in programming that is designed for helping humans live together.[150]

"In the beginning God created the heavens and the earth" (Genesis 1:1). This first chapter of Genesis opens with God speaking light and more into existence. The first chapter of John's Gospel begins in a similar way. "In the beginning was the Word, and the Word was with God, and the Word was God" (John 1:1). Speaking and listening have always been important aspects of connecting to God's truth. Communication is vital to knowing Him and understanding our place in His world.

Looking at the Eyes of Other Humans:

Have you noticed that the part of another person's face that you are drawn to is their eyes? Apparently, God built into all humans the realization that you can get to know people very well by looking into their eyes.

"Newborns can typically see about a foot away, which **happens** to be the distance to a mom's eyes when breastfeeding. Around 2 months old, babies can focus and make eye contact. The older the infant gets, the more interactive and progressive the eye contact becomes"[151] (emphasis added). Not surprisingly, a study shows that eye contact between baby and mother increases a chemical called oxytocin which seems to increase bonding.[152]

"(S)cientists have already established that our brains are **built to communicate** with each other. Previous research has shown that when babies and caregivers **look into each other's eyes, they bond**" (emphasis added). This research showed that there was more brain wave activity in both the baby and the adult.[153]

It is all just more proof that God is in control of the details of our lives.

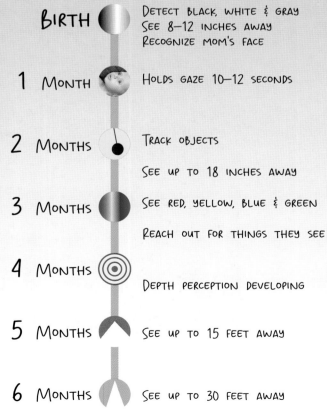

Eye contact, especially between a child and mother, triggers a neural pathway that leads to the release of oxytocin. It is believed that certain brain regions involved in social bonding, such as the amygdala and the hypothalamus, are activated by eye contact.

Infant Vision Development Timeline

BIRTH
Detect black, white & gray
See 8–12 inches away
Recognize mom's face

1 Month
Holds gaze 10–12 seconds

2 Months
Track objects
See up to 18 inches away

3 Months
See red, yellow, blue & green
Reach out for things they see

4 Months
Depth perception developing

5 Months
See up to 15 feet away

6 Months
See up to 30 feet away

27
How long does a child have their natural language instinct?

The Bible makes it clear that mankind had only one language until about 4,500 years ago. When men disobeyed God by not spreading out across the face of the planet after Noah's Flood and instead started building the Tower of Babel, God corrected the situation by creating many different languages. This caused people to move away from each other and populate the whole Earth (Gen. 11:1–9).

Today, with our great world-wide communication systems, we come into contact with other languages regularly. Learning another language can be very difficult. However, as we previously noted, even babies have the ability to learn several languages.

This author has studied five languages other than the one he was born into, but the first extra language was at age 13. His grades were A's, but each was difficult and did not become easy to use. The studies mentioned earlier about our ability to learn new skills explain that the author's brain had been "hard-wired" to use only one language and required large "software programs" to learn each of the five extra languages.

SOME PARENTS TEACH THEIR BABY SIMPLE SIGN LANGUAGE, LIKE THE "MORE" SIGN. USING SIGNS ALLOWS YOUR BABY TO EXPRESS HIM OR HERSELF WITH GESTURES BEFORE HE OR SHE IS ABLE TO SPEAK WITH WORDS. RESEARCH SHOWS THAT THIS CAN HELP YOUR BABY LEARN TO SPEAK WITH WORDS SOONER

Dr. Steven Pinker is a language expert and professor at MIT. His studies strongly indicate that "how to learn" a language is already "hard-wired" at birth (an instinct for learning languages). His understanding explains this author's language difficulty after age 13: "By puberty, the faculty for language acquisition has deteriorated, which is why tourists and students have such a hard time learning a foreign tongue."[154]

Dr. Pinker is so sure learning a language is programmed into newborn babies that he wrote a book about it: *The Language Instinct.*[155] He has also written articles and chapters of books about it.[156] The following will be the ideas presented by Dr. Pinker.

Language as an Instinct
1   The universality of language.
2   The universality of the design of language — mental algorithms that underlie people's ability to talk.
3   Language development in children is the same way in all the world's cultures.
4   Language seems to have neurological and perhaps even genetic specificity.

"**Language is a complex, specialized skill,** which develops in the child spontaneously **without conscious effort or formal instruction,** is deployed without awareness of its underlying logic, is qualitatively the same in every individual, and is distinct from more general abilities to process information or behave intelligently… All this suggests that language is caused by **dedicated circuitry**… Why do I call language an instinct? There's no such thing as a Stone Age language.

Those are the first **two bits of evidence, the universality of language and the universality of the design of language** — that is, the kinds of **mental algorithms that underlie people's ability to talk.** The **third bit of evidence** is from my own professional specialty, **language development in children.** We see language development proceed **the same way in all the world's cultures.** It's remarkably rapid, as any parent can attest.

And what the child has done is **solve a remarkably difficult computational problem… that's what the child does in those six months,** despite the lack of grammar lessons or even feedback from parents… The **final bit of evidence** is that language seems to have **neurological** and perhaps even **genetic specificity**"[157] (emphasis added).

"(O)n the left side of the forebrain… is **a complex set of neural circuits that have been programmed with 'super-rules'** (making up what Mr. Chomsky calls '**universal grammar**'), and that these rules are UNCONSCIOUS AND INSTINCTIVE… (T)he unborn fetus must already react to language, for experiments with 4-day-old French infants show that… (T)hese INBORN LINGUISTIC **mechanisms** really take off at about 18 months after birth, and they are fully operating at about 3 years, when **even the errors a child commits follow precise grammatical rules.** By puberty, the faculty for language acquisition has deteriorated, which is why tourists and students have such a hard time learning a foreign tongue…"[158] (emphasis added).

THE MOST COMMON FIRST WORD IS "DADA"

Dr. Pinker sees the ability of a six-month-old baby to "solve a remarkably difficult computational problem" because of "a complex set of neural circuits that have been programmed with 'super-rules.' " Yet the doctor does what ROMANS 1:18–20 says is done by people who reject God: he gives the credit to evolution (no Designer made this).

28
Why don't we see black when we blink?

"From crying to spitting out items that hit their tongues, **babies come out** of the womb **ready** to do what they can **to protect themselves…** Realizing **the genius** of how babies are **designed** is fascinating for new parents. We already know that our kids are **miracles,** but understanding how much they automatically know how to do at birth helps us see even more **how uniquely they are designed"**[159] (emphasis added).

This means that the instincts necessary for protection are already present and working when the child is born. We've already discussed many instincts necessary for staying alive (like breathing) and for a healthy life. Now let's look at some additional ones necessary for protection.

Blinking: By now you should be familiar with the idea that almost nothing that we see around us in nature is as simple as you thought. Everything is complex and marvelously made! That is also true of the "simple" act of blinking.

God made an instinct (some call it a reflex) to have the eyelids open and close at a rate of once every few seconds, and this rate can change depending on need. The blinking protects the eye in routine ways and in emergencies.

The routine protection is what happens every time you blink. "Blinking is essential for your eye health."[160] As the eyelids close, they spread a mixture of oils and mucous secretions across the surface of the eye to keep it from drying out.[161] The moving of these "tears" do at least the following: clear debris from the eye, bring nutrients to the eye, keep the eye wet, bring oxygen to the eye, and help prevent eye infections.

Most people do not realize how important this instinct is. If humans do not blink enough the following can happen: eye pain or blurry vision, the cornea (directly over the iris) can swell, or get eye infection from debris or lack of oxygen or nutrients.[162]

The hearing ear and the seeing eye,
The LORD has made them both.
(Proverbs 20:12)

Blinking can also prevent harm being done to the eye. Sudden exposure to very bright lights will cause blinking. If something like dust contacts the eyelashes, you will blink. Your instincts will quickly close the eyelids if anything is quickly coming at your face.

Remember that even a baby has all this protection.

Have you ever considered the fact that you seldom notice your blinking? Your eyes are closed every few seconds, so why don't you see darkness every few seconds? This amazing instinct has a complex method of ignoring the momentary blackouts. "The very act of blinking suppresses activity in several areas of the brain responsible for detecting environmental changes, so that you experience the world as continuous."[163] This is an important benefit that you probably never knew you had: almost never seeing the many blackouts that occur every minute of your waking life.

? Since we blink many times every minute, why don't we see black when we blink?

Diving Instinct: Babies are born with a diving reflex (instinct). When cool water hits the face, the body prepares to go under water. Few people know about this protection God built into baby brains, partly because it disappears after the first half year of life outside the womb.[164]

If a child falls into the water, several things will change in his body to help him survive. When the face is exposed to water below 70°F, the following actions are done by the brain. First, the baby will hold his breath. Blood vessels to the body's extremities will constrict to reduce loss of heat. This will also make the blood "mainly just move between the brain and heart so the vital organs are taken care of."[165] The heart rate slows down by 10–25% to maximize oxygen output.

If the child goes extremely deep under water, "the **body intentionally** allows" fluid to fill the lungs and chest cavity. This will prevent the organs from being crushed by the great pressure.

The writer's use of the words "body intentionally" is another way of saying that this is controlled by the unconscious mind, not the conscious mind. This is confirmed by the fact that the above list of amazing, protective actions will happen even if the individual is unconscious before entering the water![166]

Parachute Reflex: A very important, protective reflex is called the parachute reflex. It appears when a baby is about 6 months old, and it never goes away. The name probably comes from the fact that when someone starts to fall, the arms and legs extend "on both sides of their body in a protective movement, exhibiting a SKYDIVING like posture… Even though this reflex appears long before babies walk, the parachute reflex's **primary purpose** is to break a baby's fall and is **needed** for all the practicing and subsequent falls that will ensue whilst learning to walk"[167] (emphasis added). So, God added this reflex just before a baby would start falling while trying to walk and run. It's another nice design by God.

He also decided to leave this instinct in our bodies for the rest of our lives. As one scientist said, "the parachute reflex stays with us throughout adulthood, as a handy security mechanism. Do you remember the last time you face planted the floor? 'nuff said!"[168]

Withdrawal Reflex: Babies are often unable to know whether something coming at them might hurt them. So God put into them the withdrawal instinct. If anything is coming quickly at a baby's face, it will instinctively turn its head to the side. This can confuse a mother who is moving her head toward her baby to kiss him, and he turns away.

As instincts go, this one is relatively simple. If a toy or blanket is moving toward his face rapidly, he will turn his head to avoid impact. This reflex will never go away, but the child will learn when it is actually necessary to withdraw. He will learn the difference between a baseball flying at his face and his mother moving in to wipe peanut butter from his cheek.[169]

? What are the amazing changes caused by instinct in a baby's body to help save his life, if he falls into a lake?

29

What is so incredible about the simple act of a baby crying?

A newborn baby cannot control much of his body, much less talk. So how does he communicate with his mother if he has a need? Usually, it is through crying.

"Without cries to keep adults attuned to their needs, **babies wouldn't survive**... Researchers also found that an infant's cry is **naturally calibrated to affect adults,** even those who aren't the child's parent. Our brains work in a way that makes it almost impossible to ignore the cry of a baby, which is why being stuck on a plane with an upset infant is difficult"[170] (emphasis added).

Again, as this book has been pointing out, even this "simple" act is not simple.

First, some part of the baby's brain must receive and correctly interpret a signal that there's a problem in the body. Possibly there's a need for food or liquid. Possibly there's a need to expel waste from the digestive system. Possibly there is pain from a sharp object in the clothing around his body. God has put many system alarms into a baby's body.

Second, the baby's brain then needs to determine how to correct the problem. For a newborn there is usually only one instinct available: crying. The brain will trigger the crying instinct.

However, the brain will almost "talk" to the adult through the crying instinct. Mothers and doctors have learned that, with practice, they can interpret the crying to have a good idea of what the baby's problem is. Several websites have lists helping mothers understand what each kind of crying means.[171]

So the action called "crying" is not just noise. The crying instinct is controlling the baby's diaphragm, lungs, vocal cords, mouth, etc., to almost talk to the adult. The online lists encourage the mother to listen and watch for things like volume, pitch, rhythm, repetition, sucking motion, lip smacking, putting fingers to the mouth, nasal sounds, turning the head, turning the body, whimpering, screaming, etc.[172] Think about it. This instinct is incredible considering it has to use the very undeveloped body and brain of a newborn baby!

	Newborn Crying Code	
1	Hunger	Fairly desperate and unrelenting; usually high pitched.
2	Tired	Breathy, helpless. This cry can be intermittent and is more easily soothed than others.
3	Boredom or overstimulation	Usually not as loud as other cries, and often staccato. Boredom can easily transition to laughter; overstimulation can escalate to shrieking.
4	Annoyance or Discomfort	Forced and whiny; has a pattern of short repetitions, like "uh-UH, uh-UH."
5	Pain	Piercing and grating.

1	2	3	4
It controls the voluntary and involuntary movement of the body.	It maintains the balance of the body.	It maintains the homeostasis of the body by doing functions such as thermal regulation, etc.	It is responsible for sensing the stimuli, processing them, and sending the appropriate response.

Crying from Pain: The second chapter of a great book was titled, "The Gift No One Wants." That gift is pain! The chapter explained in detail the marvelous system God has put into humans, even babies, that protects us in many ways. The chapter was so good that Dr. Paul Brand and Philip Yancey turned it into a whole book: *Pain: The Gift Nobody Wants.*[173]

Without pain, God's creatures, not just man, would let the cause of the pain continue doing its damage. The book by Dr. Brand pointed out that the main symptom of leprosy is the lack of pain. That is why pictures of lepers will show them missing fingers and toes; they never felt the continual damage being done to them. As the book explains, pain is like a fire alarm; it is not the fire! The alarm may hurt, but the fire is what causes the damage.

God has given this gift to even little babies.

The pain system used by our instincts uses at least FIVE KINDS OF RECEPTORS which are scattered around our bodies in the most useful places, which include nociceptors and thermoceptors. Amazingly, some of the receptors are "silent" until our body's immune system causes an inflammation (by sending extra blood to a damaged area), which releases certain chemicals that "turn on" these silent receptors and we "feel" a dull pain.[174] All of these five kinds of receptors send their signals through the nervous system.

Our nervous system uses at least four types of nerves for sending signals that are interpreted by our instincts. One type is used by our sense of balance and eyes. Another is used for our sense of touch. The instincts that detect and use pain information use at least two other types of nerves — one for sharp pain and one for dull pain. The sharp pain nerves are designed for very fast transmission of the signal, helping the instinct to react as rapidly as possible. The dull pain nerves are not as fast but allow for more sensitivity to detect weaker signals (less important problems), such as an itch.[175] NOW IF GOD HAD GIVEN ONLY PAIN WITHOUT THE INSTINCT TO USE IT, THAT WOULD BE TERRIBLE. HOWEVER, IN HIS LOVE FOR US, HE PROVIDES THE INSTINCT THAT TAKES THE PAIN AND CAUSES EVEN A BABY TO ACT TO FIX IT. The usual response of a baby is to cry, loudly. God has built into adults, especially the mother, a response of quick action to try to relieve the crying baby of the pain.

As children grow, the response to the pain is usually to get away from what seems to be the source. Once the pain has been stopped or at least reduced, the child can search for help in fixing the problem causing the pain.

5 Pain Receptors	
1	nociceptor
3	thermoceptor
4	mechanoreceptor
6	chemoreceptor
7	polymodal

30
How do instincts prepare a mother to nourish her baby?

One of the most important parts of life is the care of a baby. The baby is almost completely helpless. So he needs to feel and actually receive nourishment, love, and security. God has provided these mainly through instincts.

Love and Bonding: God made the newborn to have a "preference for the mother's voice rather than another human voice."[176] Besides being attracted to his mother's voice, he is also attracted to her face, at least partly because that is the face he sees while getting the delicious mother's milk. A newborn is only able to see about 18 inches, but that is the distance to his mother's face when he is breastfeeding.[177]

As mentioned earlier, the fact that looking at the eyes causes better parent-child bonding. Scientists have discovered "eye contact increases a brain chemical called oxytocin, the very same chemical present in parent-child bonding."[178]

From the mother's standpoint, God has put instincts inside her to take care of children, especially her own. Parental love in the mother is very strong, especially during infancy and early childhood, but it usually remains to the end. It is "unquestionably instinctive in the mother ... ONE OF THE MAJESTIC FORCES IN THE HISTORY OF HUMANITY"[179] (emphasis added).

Nourishment: Mother's milk has everything a baby needs to eat. The baby's instincts drive him to search for the mother's breast and begin sucking out and swallowing the milk. As usual, this instinct of the baby is anything but simple.

The mother's instincts will have prepared her body to provide that milk. In the first few days after birth, before regular milk is produced, her body provides the newborn with a sticky, yellowish fluid sometimes called "liquid gold" because it is so great for the baby. Then the regular milk will start flowing at a rate that would satisfy several babies. This rate will adjust over the days for the actual number of babies. Plus, since the mother's milk is affected by what she eats, the baby is exposed to many different tastes, making transition from mother's milk to other foods easier.[180]

Mother Willing to Die for HER Child:

"While generally humans and other animals flee or freeze when faced with an imminent threat — mothers stay put to protect their babies." Neuroscientists have proven that a major factor in this change in behavior is the hormone oxytocin.[181] At least two studies with rats have shown this.

One rat study found that virgin lab rats, when they heard cries of newborn rats, would either ignore them or possibly eat them! When the scientists put mother lab rats in the same situation, they would protect and care for the newborn rats. Then the scientists injected the virgin lab rats with oxytocin and repeated the situation of them hearing the newborn rats crying: instead of eating them, the virgin lab rats protected and cared for the newborn rats![182]

Another study put mother rats in a dangerous situation, with and without their newborns. Alone, they froze from fear. However, with their newborns next to them, they attacked the source of the threat! Then the scientists blocked the oxytocin's effects in the mothers and put them in danger with their newborns: this time the mothers froze from fear.

So oxytocin is a major factor in the maternal instinct that causes mothers who see danger for their child to say things like the following. "I realised that if he came anywhere near my child, I would have killed him rather than let him harm my baby … I have no doubt that if tested, my strength would have been superhuman in the face of the threat to my baby."[183]

? Why are mothers willing to fight and even die for their own child?

OXYTOCIN, OFTEN REFERRED TO AS THE "BONDING HORMONE," FOSTERS A SENSE OF CONNECTION AND STRENGTHENS EMOTIONAL BONDS BETWEEN INDIVIDUALS.

Amazing Oxytocin!		
ONGOING RESEARCH IS CONTINUALLY EXPANDING OUR UNDERSTANDING OF THIS AMAZING HORMONE AND ITS IMPLICATIONS FOR HUMAN BEHAVIOR AND WELL-BEING.	Role in Labor and Breastfeeding	Oxytocin plays a crucial role in childbirth by stimulating uterine contractions and facilitating labor. It also aids in breastfeeding by promoting milk ejection or the "let-down" reflex, allowing the release of breast milk.
	Impact on Social Behavior	Oxytocin influences social behavior by enhancing trust, empathy, and generosity. Research suggests that it can increase prosocial behavior, cooperation, and bonding between individuals.
	Stress and Anxiety Reduction	Oxytocin has been linked to stress reduction and anxiety relief. It can help mitigate the effects of stress by reducing cortisol levels and promoting relaxation.

31

Why is the hunger instinct so crucial for survival?

Being Hungry for the Correct Food: "We couldn't have survived this long without the body's ability to feel hunger," says Joanne Slavin, PhD, RD, a professor of food science and nutrition.[184] Hunger is crucial for survival, but it is an extremely complex instinct.

What is the correct food for humans? It would be what keeps us the healthiest. Studying the eating habits of Americans and people from around the world, food experts have been able to determine what works best in human bodies. A simple summary would be to eat food the way God made it.[185] Avoid processed foods as much as possible. Whether the food is processed or not, the hunger instinct still works.

This instinct is such a multiplex that only a simplified explanation can be given here. Many different parts of the human body are communicating with several parts of the brain. Various chemicals are being put into the blood stream or stomach (mostly hormones for instruction and enzymes for digesting). The body adjusts to whatever kinds of food it receives. The following is a simple presentation of being hungry, eating, and feeling satiated (satisfied).[186]

Several hours after eating, your stomach is empty, and the hypothalamus (a small part of the brain) gets signals from hormones from "various parts of the body that deal with energy intake and storage…."[187] The hypothalamus responds by producing two proteins that make you feel hungry.

Next you eat enough food to satisfy that feeling of hunger. There are SEVERAL WAYS THE FEELING IS SATISFIED. The most common ways are BLOOD SUGAR LEVEL, AMOUNT OF PROTEIN, AMOUNT OF FIBER, AND SIZE OF THE MEAL.

When the desired BLOOD SUGAR LEVEL is reached, the pancreas will add insulin to stabilize it. The pancreas will signal the hypothalamus, which will produce two different proteins that make you feel satisfied.

One result of eating most processed foods is a spike in the blood sugar level. The refined sugar gets into the blood stream too quickly. The pancreas reacts to the sudden increase as if it was caused by a large amount of "normal" food (unprocessed): it adds too much insulin for the actual amount of sugar that will be coming. The excess insulin soon causes the blood sugar level to drop too low, making your hypothalamus conclude you need more food. The result of adding processed foods, especially something like a candy bar, to an empty stomach is that you are quickly hungry again![188]

As humans grow up, God has apparently allowed them more freedom of choice in response to issues that arise. Adult humans seem to be able to override many of the instincts in their bodies. However, they still have many instincts necessary for a healthy life.

? Why is eating a candy bar when you have an empty stomach a bad idea?

Another way to stop being hungry is with protein (lean meat, seafood, milk, etc.). When enough protein is detected in the stomach, "fullness hormones" are released. The hypothalamus detects them and produces the proteins for the feeling of satisfaction.

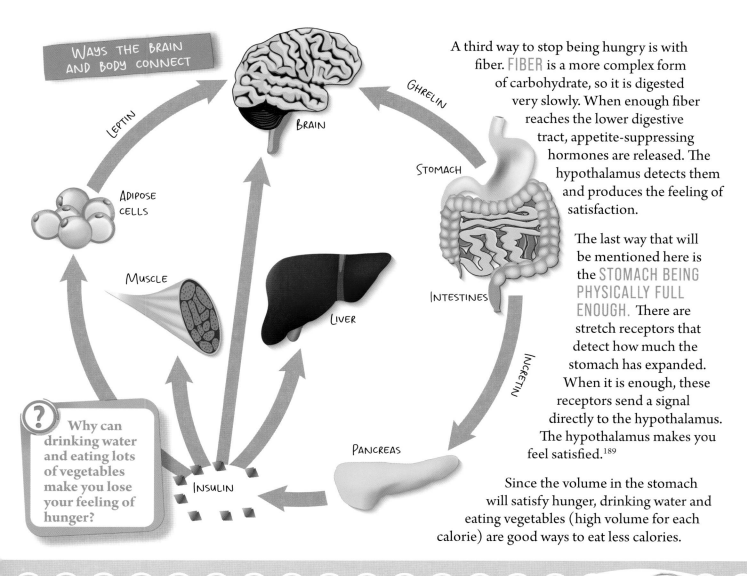

LEPTIN

GHRELIN

BRAIN

STOMACH

ADIPOSE CELLS

MUSCLE

LIVER

INTESTINES

INCRETIN

PANCREAS

INSULIN

? Why can drinking water and eating lots of vegetables make you lose your feeling of hunger?

A third way to stop being hungry is with fiber. FIBER is a more complex form of carbohydrate, so it is digested very slowly. When enough fiber reaches the lower digestive tract, appetite-suppressing hormones are released. The hypothalamus detects them and produces the feeling of satisfaction.

The last way that will be mentioned here is the STOMACH BEING PHYSICALLY FULL ENOUGH. There are stretch receptors that detect how much the stomach has expanded. When it is enough, these receptors send a signal directly to the hypothalamus. The hypothalamus makes you feel satisfied.[189]

Since the volume in the stomach will satisfy hunger, drinking water and eating vegetables (high volume for each calorie) are good ways to eat less calories.

One Reason Americans Eat Too Much: Unfortunately, in modern American society, most of our food has been processed. Food the way God designed it works well with our bodies. However, manufacturers have realized that salt and sugar are key ingredients used by our tasting instinct to determine whether something is good for our bodies. In unprocessed foods, there are plenty of vitamins and minerals for every little bit of sugar or salt. In processed foods, that is not the case.

Generally, the more processing of a food (changing it from the way God made it), the more vitamins and minerals are lost. Even when manufacturers "enrich" or add vitamins and minerals to a product, studies show it's not as good as before processing. Processed food "can contain as much as 8 times the amount of sugar as whole foods…" We eat but do not get much nourishment. Our other instincts tell us we need more nourishment, so we keep eating. That is a major reason modern Americans are more overweight and unhealthy than most Europeans.[190, 191, 192]

32

What is involved in the response to touching something hot?

Pulling Away from Heat: You've probably accidentally touched something hot and reacted very quickly by pulling your hand away. You may not know it, but you moved your hand before your conscious mind was able to understand what had happened!

This super-fast reaction is because of "the **special ability** that **evolution gifted us** to facilitate our survival"[193] (emphasis added). Of course this writer did not even try to explain HOW evolution (dumb luck) "gifted us" with this marvelous protection. Unfortunately, as is common with evolutionist writers, fellow evolutionists have published many of his articles. Writers who see and report the idea that God gave us this gift have a difficult time finding a publisher.

Let's look at this "special ability" God gave us.

When you touch a very hot pot, the skin receptors quickly send electrical impulses along sensory neurons to the spinal cord. In the spinal cord, the impulses are processed. The spinal cord is where this instinct amazingly decides what muscles need to contract and not contract to prevent more damage. Once the decision is made, the spinal cord's interneurons send the correct electrical impulses to the correct motor neurons. Those motor neurons stimulate the correct muscles to pull your hand away from the hot pot.

This is not simple. Just to pull the hand away quickly requires at least signaling the biceps to contract and the triceps (on the opposite side of the arm) to relax and let the biceps easily move the arm away from the heat. The instinct does more than that. It has the spinal cord also signal the rest of the body to adjust to the sudden movement of the affected arm, so the body doesn't lose its balance and possibly fall over.[194]

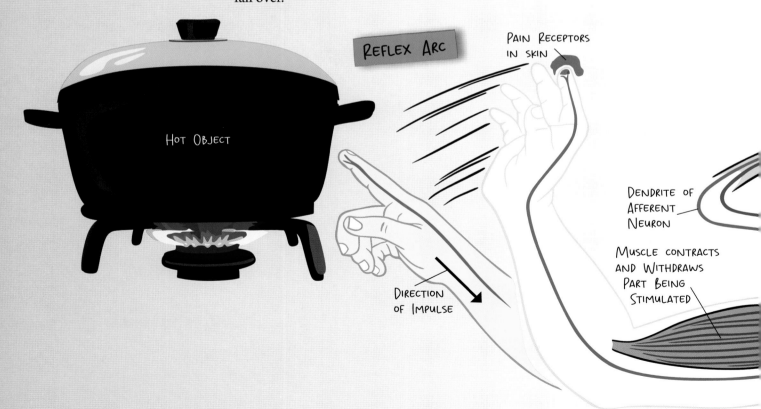

REFLEX ARC

PAIN RECEPTORS IN SKIN

HOT OBJECT

DIRECTION OF IMPULSE

DENDRITE OF AFFERENT NEURON

MUSCLE CONTRACTS AND WITHDRAWS PART BEING STIMULATED

To better understand how complex this instinct is, think about the fact that the spinal cord can get a sharp pain signal from almost anywhere on the body's surface. It then will need to correctly choose which muscles to contract and which to relax, in order to move your body away from the source of pain. It, at the same instant, will need to choose which muscles to contract and which to relax on the opposite side of the body, in order to keep the body in balance and not possibly fall down. From your own experience, you know the pulling away part of the instinct works well. But one scientist said the balancing part of this reflex also works very well, "virtually ensuring that the opposite limb provides stabilization."[195]

"This process happens so fast that the response occurs before the message reaches the brain."[196] Now that is fast! But it needs to be fast to prevent significant damage. Waiting for the conscious brain (the cerebrum) to process the signals and decide what to do "would take too long, potentially exposing the body to risks."[197]

In 1649, René Descartes, a French philosopher and scientist, put forward the initial mechanistic explanation for what would later be known as the Reflex Arc. Descartes suggested that the body's actions are reflexive, while the mind's actions are deliberate, conscious, and meaningful. His proposition relied on a simple model where sensory information is received and processed by the brain, leading to motor responses. However, at that time, the understanding of neural activity was lacking, which compelled Descartes to explain it as being driven by "animal spirits."

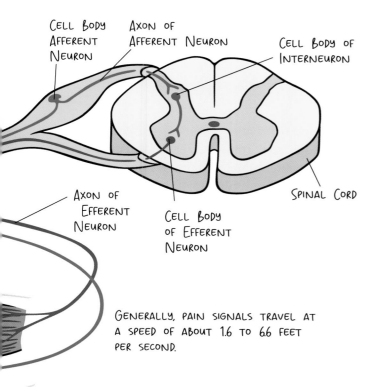

CELL BODY AFFERENT NEURON

AXON OF AFFERENT NEURON

CELL BODY OF INTERNEURON

AXON OF EFFERENT NEURON

CELL BODY OF EFFERENT NEURON

SPINAL CORD

GENERALLY, PAIN SIGNALS TRAVEL AT A SPEED OF ABOUT 1.6 TO 6.6 FEET PER SECOND.

? **How fast do pain signals travel through your body?**

As mentioned earlier, SMELL is a very important instinct for guiding humans about food. It is great for encouraging us to eat things that are good for our bodies and avoid things that are bad. We'll now look at how it handles bad odors.

Good vs. Bad Odors: For protection, the nose is very important. The human nose can detect about one trillion odors.[198] God gave the brain the ability to distinguish with very good accuracy between what is good for the human body and what needs to be avoided. THIS INSTINCT IS EVEN ABLE TO BE FAIRLY ACCURATE ABOUT THE DEGREE OF GOODNESS AND THE DEGREE OF HARMFULNESS.

Think about the last time you smelled something bad. Most likely your desire to get away from that smell was an appropriate response to the amount of harm that it could have done to you.

Scientists are still trying to understand how our instincts judge this huge number of scents. One team studied over 1,500 properties of about 150 molecules that humans judge to range from very good to very bad smell. The result seemed to show a correlation to molecular size and electron density. However, when they tested their theory, it failed almost 70% of the time![199] In other words, the amazing smell instinct is still amazing to scientists.

Bad Odors and Pain Turned Off Over Time: Another great behavior of the human smell and pain instincts is that they turn off the warning after the conscious brain has had sufficient time to respond. Remember that a bad odor is not really the problem, just like pain is not really the problem. Both bad odors and pain are warnings of something that could harm the body. They are there to get the conscious mind to fix a problem.

If the smell or pain warning is of a significant problem, it may lessen in intensity, but it will continue until resolved. For a terrible smell, the source of the smell needs to be prevented from sending its molecules to your body. For a sharp pain, the cause of damage to your body must be removed. If significant damage has already been done, then healing must occur.

However, if the smell or pain warning is of a minor problem, God has designed it to turn completely off after a few minutes. This is the reason you can be using the toilet and be surprised when someone comes into the restroom and says you stink! You're surprised because after a few minutes of the smell, the instinct determined you had been warned enough, so it turned off the warning. The same thing, fortunately, happens when you get into a hot tub; it goes from very hot to comfortable within a short time. "These changes are neural adaptation and they are useful most of the time."[200]

"This whole process is pretty intense for your brain. To keep your nervous system from exhausting itself with continuous stimuli, the receptors experience temporary sensory fatigue, or olfactory adaptation. Odor receptors stop sending messages to the brain about a lingering odor after a few minutes and instead focus on novel (new) smells."[201]

> **?** Why do you stop noticing a bad smell after a few minutes? Is this okay?

To better understand how important this instinct has been to the human race, think about what "bad" smells have been "lived with" by numerous civilizations. Before the "civilizations" realized they should obey the Old Testament commands to bury their excrement, wash in running water, and not touch dead animals (mainly after Lister and Pasteur warned the world of invisible microbes in the late 1800s), there were many odors that were mainly warnings of harmful chemicals or microbes, and people ignored them.

Before automobiles, horses were abundant everywhere. The horse manure was routinely scooped up and usually piled on street corners until finally hauled out of town. People would seldom bother to remove even their own excrement from the streets. It was common in Europe to yell a warning and throw what now goes down our toilets out into the street! Another bad odor "civilizations" lived with was caused by people seldom bathing; we laugh when we learn that few bathed even

once a month. Also, in some civilized places, the dead bodies of animals, especially small ones, were not quickly disposed of.

The Black Death, or the bubonic plague, of the 1300s that killed about a third of the people in Europe (more than 25 million) was caused mainly by fleas on rats.[202] They thrived in the filth the Europeans didn't dispose of.

However, the Jews living in Europe generally followed the guidance of the Old Testament, doing things like burying excrement and washing in running water. The result was that "Jews were perceived as being less susceptible to the plague than their neighbours (likely the result of Jewish ritual regarding personal hygiene…"[203] It's another example of the wisdom God put into the Bible; follow it and you'll avoid many problems in this life.

Health Guidance in the Old Testament		
Dietary laws:	Leviticus 11:1–47	outlines the various types of animals that are allowed and forbidden for consumption according to Jewish dietary laws.
	Deuteronomy 14:3–21	provides a similar list of forbidden and allowed animals, along with instructions on how to slaughter animals for consumption.
	Exodus 23:19 and Exodus 34:26	prohibit boiling a young goat in its mother's milk, which is interpreted as a prohibition against mixing meat and dairy products.
Quarantine laws:	Leviticus 13:1–59	contains instructions for identifying and isolating people with various skin diseases, such as leprosy.
	Leviticus 15:1–33	contains instructions for isolating people with certain bodily discharges.
Hygiene laws:	Leviticus 11:32–40 and Numbers 19:11-22	contain instructions for disposing of contaminated objects and for purifying oneself after coming into contact with the dead.
	Deuteronomy 23:12–13	contains instructions for keeping the camp clean by disposing of human waste outside the camp.
Sabbath observance:	Exodus 20:8–11	includes the commandment to observe the Sabbath day as a day of rest and worship.
	Leviticus 23:3	instructs the Jewish people to observe the Sabbath day as a holy convocation, a day of rest and worship.

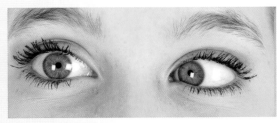

34

What instinct is needed to keep your eyes from being endlessly blurry?

VOR to Prevent Blurring: You've probably heard of drunks getting blurry vision. That is because the ethyl alcohol they drank messed up the marvelous system God gave us to keep our eyes locked onto what we're looking at. The system is called the Vestibulo-Ocular Reflex (VOR) and its main job is to "stabilize images on the retina… The VOR is vital because people are constantly making small head movements."[204] Without this instinct, your vision would be blurry most of the time.

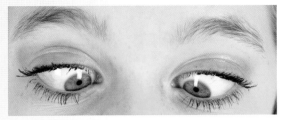

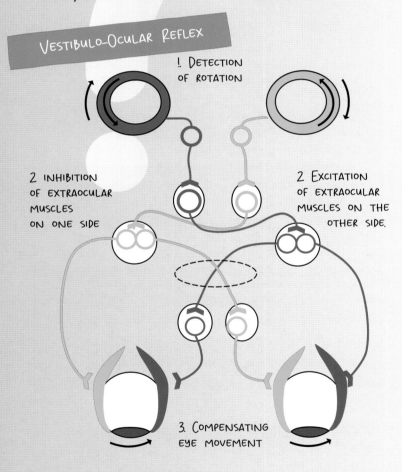

VESTIBULO-OCULAR REFLEX

1. DETECTION OF ROTATION

2. INHIBITION OF EXTRAOCULAR MUSCLES ON ONE SIDE

2. EXCITATION OF EXTRAOCULAR MUSCLES ON THE OTHER SIDE.

3. COMPENSATING EYE MOVEMENT

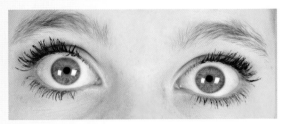

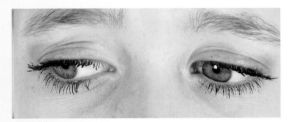

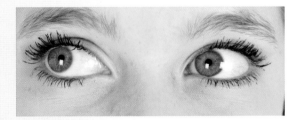

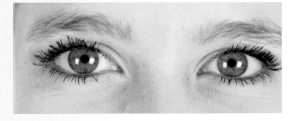

The pons in the brain stem is the main controller of this instinct.[205] Somehow it takes the head movements planned by the rest of the brain and orders both eyes to move in the opposite direction and the same distance! Think about how often and in what directions you move your head. Move your head right now while looking at something small. Move your head in any direction. Amazingly, that small object will stay in focus as the pons makes both of your eyes move the correct amount at the right speed. This reflex is one of the fastest in the human body. "It has been estimated that the eye movements lag the head movements by less than 10 milliseconds."[206] Remember that there are 1,000 milliseconds in just one second.

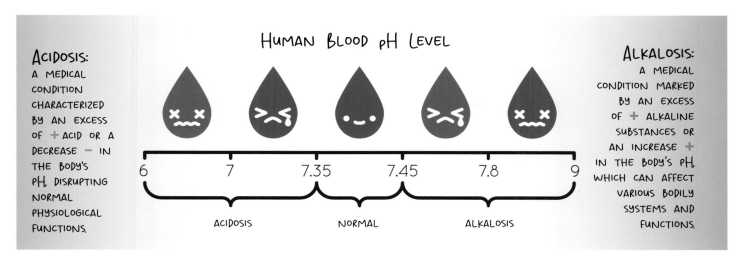

HUMAN BLOOD pH LEVEL

6   7   7.35   7.45   7.8   9

ACIDOSIS   NORMAL   ALKALOSIS

ALKALOSIS: A MEDICAL CONDITION MARKED BY AN EXCESS OF + ALKALINE SUBSTANCES OR AN INCREASE + IN THE BODY'S pH, WHICH CAN AFFECT VARIOUS BODILY SYSTEMS AND FUNCTIONS.

System Stability (Homeostasis): "Homeostasis is the process in which the body maintains normal, healthy ranges for factors such as temperature, energy intake and growth. Homeostasis is essential for healthy bodies."[207] Your conscious mind barely knows what's going on. Your unconscious mind is constantly watching and adjusting the various conditions that are important for good health.

As mentioned earlier, temperature is key for the huge number of chemical reactions going on all the time in your body. The pH is also important for those chemical reactions. The need for the correct amount of water in your body has already been discussed. The amount of sugar in your blood also needs to be controlled.

All of this control is made more difficult because you're not dead! You are continually using your body and changing all of the crucial factors (temperature, pH, water, sugar, etc.). Thus, the controlling of them is difficult but "essential for healthy bodies."[208]

HOMEOSTASIS IS THE BODY'S ABILITY TO MAINTAIN A STABLE INTERNAL ENVIRONMENT BY REGULATING VARIOUS PHYSIOLOGICAL PROCESSES AND BALANCING FACTORS SUCH AS TEMPERATURE, pH, AND NUTRIENT LEVELS.

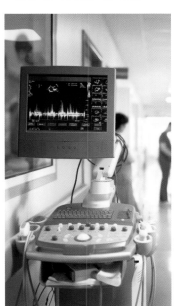

Vital Signs-Adult	Normal Level
Temperature	97.8°F to 99.1°F (36.5°C to 37.3°C); average 98.6°F (37°C)
Heart Rate (Pluse)	60 to 100 beats per minute
Blood Pressure	90/60 mm Hg to 120/80 mm HG
Respiratory Rate	12 to 18 breaths per minute
Oxygen Saturation	96 to 100%
pH	7.3 to 7.5
Blood Sugar	90 to 99 mg/dL (fasting)
Proteins	150 mg per day
Electrolytes	Sodium: 136 to 144 mmol/L
	Potassium: 3.7 to 5.1 mmol/L
	Calcium: 8.5 to 10.2 mg/dL

35
Why is it your body's instinct to sweat when you exercise?

You know your body temperature should be 98.6°F. (or 37.0°C.). Your built-in instincts know that at this temperature the body's enzymes work best.[209] Controlling the temperature is done by the hypothalamus in the brain. It receives temperature signals from all over the body, considers the locations, and decides how to keep each location at the correct temperature. Engineers label the method used by the instinct as "negative feedback control."[210] We'll now look at some of the ways this instinct controls your temperature.

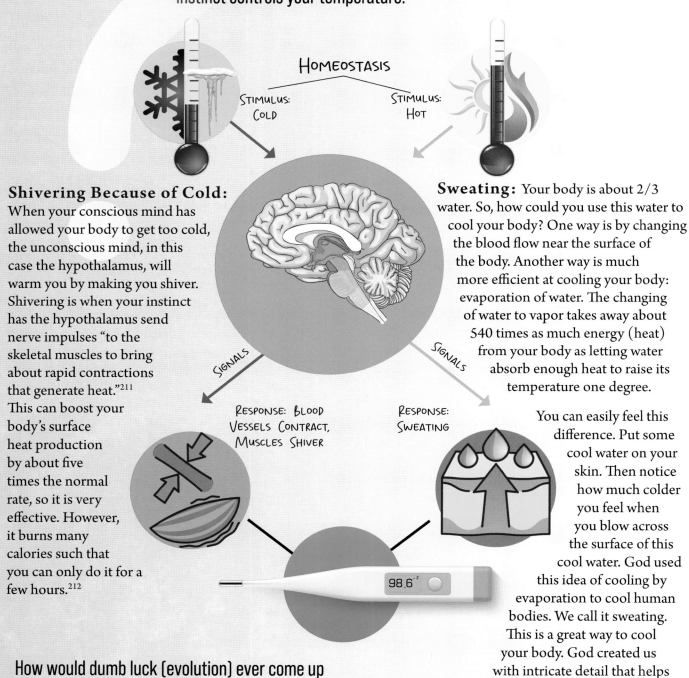

HOMEOSTASIS

STIMULUS: COLD

STIMULUS: HOT

SIGNALS

SIGNALS

RESPONSE: BLOOD VESSELS CONTRACT, MUSCLES SHIVER

RESPONSE: SWEATING

98.6°F

Shivering Because of Cold:
When your conscious mind has allowed your body to get too cold, the unconscious mind, in this case the hypothalamus, will warm you by making you shiver. Shivering is when your instinct has the hypothalamus send nerve impulses "to the skeletal muscles to bring about rapid contractions that generate heat."[211] This can boost your body's surface heat production by about five times the normal rate, so it is very effective. However, it burns many calories such that you can only do it for a few hours.[212]

Sweating: Your body is about 2/3 water. So, how could you use this water to cool your body? One way is by changing the blood flow near the surface of the body. Another way is much more efficient at cooling your body: evaporation of water. The changing of water to vapor takes away about 540 times as much energy (heat) from your body as letting water absorb enough heat to raise its temperature one degree.

You can easily feel this difference. Put some cool water on your skin. Then notice how much colder you feel when you blow across the surface of this cool water. God used this idea of cooling by evaporation to cool human bodies. We call it sweating. This is a great way to cool your body. God created us with intricate detail that helps keep our body temperature self regulated.

How would dumb luck (evolution) ever come up with such an efficient system?

Your instinct makes you sweat when your temperature sensors signal the brain that you're getting too hot. For example, you sweat when you're exercising, because the muscles are overheating and need cooling. Your instinct causes your sweat glands to cause the sweat liquid to flow out onto the surface of your skin and evaporate.[213]

As usual, God made this instinct useful for more than one purpose. A scientist wrote that the "science behind perspiration — **like basically everything that happens in our weird and wonderful bodies** — is way more interesting than you might have thought"[214] (emphasis added). Believe it or not, you perspire (SWEAT) about a quart of liquid every day through about 3 million sweat glands.[215] It is not pure water. It contains material that your body needs to expel. Sweat is a water solution with about 1–2% being ingredients like salt, protein, fats, and acids. These ingredients also include chemicals that do things like attract the opposite sex.

You may be surprised to learn that sweat is odorless. The "sweat smell" comes from those extra ingredients being quickly eaten by microbes on your skin and converted into those bad smells.[216]

Blood Vessel Size Change: Your blood is kept at the correct temperature, and it flows almost everywhere in the body. Therefore, it is very logical to use it to adjust body temperature. Since the blood flows just under the skin, it will either gain heat from or lose it to your surroundings. This gain or loss can be minimized by changing the amount of blood flow near the skin. If the body is getting too hot from the surrounding heat, the hypothalamus will cause more blood to be near the surface (to lose more heat). If the body is getting too cold from the surroundings, the instinct will reduce the blood flow near the surface (to lose less heat).

As you should expect, this process is somewhat complex. First, the instinct needs to deal mainly with the areas where the temperature is out of balance. Then, if more blood is to flow near the surface, the instinct will enlarge those blood vessels. However, to be more efficient, it will also shrink some other vessels that are relatively far from the surface. If less blood is to flow near the surface, the reverse is done: shrink the vessels near the surface and enlarge the ones far below. It's a very nice system.[217]

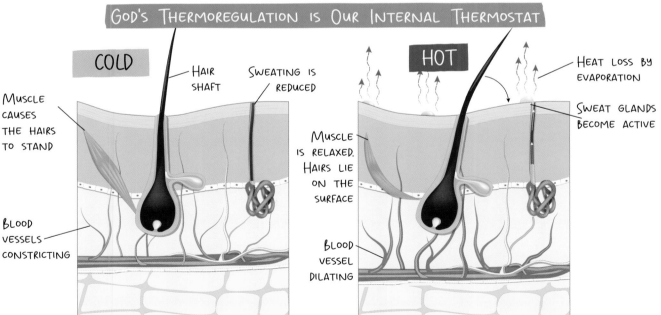

God's Thermoregulation is Our Internal Thermostat

COLD

HAIR SHAFT

SWEATING IS REDUCED

MUSCLE CAUSES THE HAIRS TO STAND

BLOOD VESSELS CONSTRICTING

HOT

HEAT LOSS BY EVAPORATION

SWEAT GLANDS BECOME ACTIVE

MUSCLE IS RELAXED. HAIRS LIE ON THE SURFACE

BLOOD VESSEL DILATING

Hair Standing on End: Another way your instincts control your body temperature uses the short hairs on your body. A way that you lose heat is by conduction as cool air blows past warm skin. If the short hairs stand up, they will slow down the movement of the air in direct contact with the skin. This would then be like a layer of insulation. If the short hairs lie down, they will allow the cool air to get closer to the skin.

God designed the short hairs with muscles attached to the bottoms, just below the skin. These allow the instinct to move the hairs from lying down to standing straight up. See the figures above for how the muscles move the hairs.[218]

36
Why is control of blood sugar in the body so crucial for your health?

EXERCISE

Control of Blood Sugar Level: Since sugar (glucose form of it) in your blood is the main source of energy for your body, a low level of it will make you weak, among other things. However, you may not realize that too much sugar in your blood can cause problems. Your instinct does know that. God designed it to regulate your blood sugar to keep it at the best concentration for your overall health.

If there is too much sugar in your blood, you can get symptoms like increased thirst, blurred vision, fatigue, and headaches. If the level gets higher, you may feel nausea, vomit, have shortness of breath, have confusion, or even go into a coma. So, this instinct is crucial for your health.[219]

The method God chose uses two hormones, secreted by the pancreas, called GLUCAGON and INSULIN. The control system, as usual, is not simple. This is how the hypothalamus controls the blood sugar level.

If your blood sugar level is only slightly low, the hypothalamus will signal the liver to convert more of its stored glycogen to glucose (the form of sugar used by your body's cells). If the level gets significantly low, the hypothalamus signals the pancreas to release glucagon, the hormone. This will go to the liver and cause it to stop absorbing glucose from the blood and start creating glucose.

The glucagon can get the liver to create glucose at least 3 ways
1. Convert stored GLYCOGEN.
2. Convert AMINO ACIDS.
3. Convert TRIGLYCERIDES (most of the fat stored in your body are triglycerides).[220]

If the blood sugar level is too high, the other important hormone, insulin, is produced by the pancreas. It will go to the liver and cause it to take more glucose out of the blood, convert it to insoluble glycogen, and store it.

"Blood sugar control involves a **complex system** of hormones," mainly glycagon and insulin, which "counterbalance each other…"[221] (emphasis added).

Finally, to reinforce the point that God did a marvelous job creating this instinct, consider the following words of a scientist summarizing the method of blood sugar level control.

"Maintenance of glucose homeostasis is **mandatory for organismal survival.** It is accomplished by **complex and coordinated interplay** between glucose detection mechanisms and multiple effector systems. The brain, in particular homeostatic regions such as the hypothalamus, plays a **crucial role in orchestrating such a highly integral response**"[222] (emphasis added).

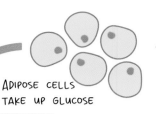

CONTROLLING BLOOD SUGAR LEVELS

ADIPOSE CELLS TAKE UP GLUCOSE

INSULIN

PANCREATIC BETA CELLS RELEASE INSULIN

INCREASING BLOOD SUGAR

DECREASING BLOOD SUGAR

PANCREATIC ALPHA CELLS RELEASE GLUCAGON

GLUCAGON

LIVER BREAKS DOWN GLYCOGEN TO GLUCOSE

EATING

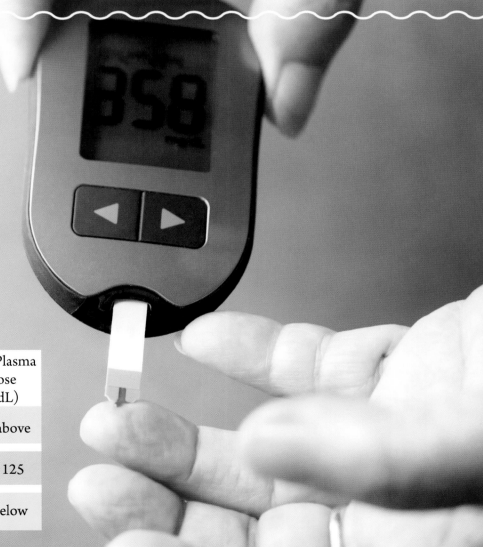

GOD HAS CREATED US WITH COUNTLESS PROCESSES THAT TAKE PLACE IN OUR BODIES EVERY DAY, OPERATING AUTOMATICALLY AND UNCONSCIOUSLY, WHICH WE MAY NOT TYPICALLY NOTICE OR ACTIVELY PERCEIVE, INCLUDING CELLULAR METABOLISM, DIGESTION, RESPIRATION, IMMUNE RESPONSES, AND NEURAL ACTIVITY.

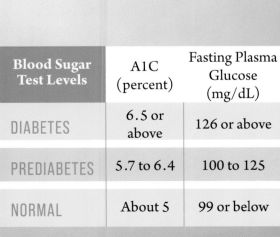

Blood Sugar Test Levels	A1C (percent)	Fasting Plasma Glucose (mg/dL)
DIABETES	6.5 or above	126 or above
PREDIABETES	5.7 to 6.4	100 to 125
NORMAL	About 5	99 or below

37
What is involved in a body's instinct to stay balanced and upright?

As we get older, we think more about our SENSE OF BALANCE. However, it is a very important instinct all through our lives. You've probably heard that the inner ear controls our sense of balance. It turns out that this instinct is very complicated and very important.

First, realize that the standing human body is very unstable. Its base (the feet) is small compared to its height,[223] and its center of mass is about halfway up its height. To show the importance of the location of the center of mass (or gravity), think about a Segway Scooter. Its handlebars are about 4 feet above the ground, but it is almost impossible to make it fall over. Its two wheels are firmly on the ground with an axle between them. That axle cannot fall. The engineers of the Segway Scooter hung the heavy lithium-ion battery under the axle. Thus, they were able to make the center of mass below the axle, forcing everything above the axle to stay above the axle.

In a human body, the feet are the only parts that cannot fall over. Since there is nothing keeping the hips (which are near the center of mass) from falling all the way to the ground, a human body is always in danger of falling over! However, God put together an amazing system to give us a sense of balance to keep us from falling over.

Reading the details of how God designed our sense of balance is very interesting. Scientists describing it use words that stress how the unconscious (not conscious) brain must continually be considering data from many sources and calculating how our muscles must respond to keep us from falling over. Here are some phrases from two writers:

"interaction of multiple systems," "process the data," "result of this central processing," "coordinated response," "(unconscious) brain interprets this information,"[224]

"integration of many different systems," "brains must rapidly and continuously integrate and then process," and "unconscious process prompts finely tuned, coordinated responses."[225]

A third researcher describing our sense of balance said, "There is **considerably more complexity** … that we have omitted…"[226] (emphasis added). So the sense of balance is definitely not simple.

Interestingly, this same writer added a very telling comment as he wondered whether one tiny part of this complex system for balance might wear out and stop working. He said, "It seems **unlikely** that a highly vulnerable system like the otoliths would be **designed** without some **method of repair**" (emphasis added). This shows that his research has taught him to expect "methods of repair" in all important parts of living creatures. He naturally uses the word "designed" without realizing he's contradicting the fundamentals of evolution.

The following is a concise description of this complex, very important system of our unconscious minds.

"Nerve fibers **from the inner ears** send balance information to the brain. Within the brainstem these nerve fibers participate in an extensive neural network involving nerves **from the eyes, the cerebellum,** and the positional receptors or 'proprioceptors' located **in the feet, legs, trunk, arms, and neck.** The **brain interprets this information, making modifications in eye, head and body position to maintain a fixed eye position, and erect posture**"[227] (emphasis added).

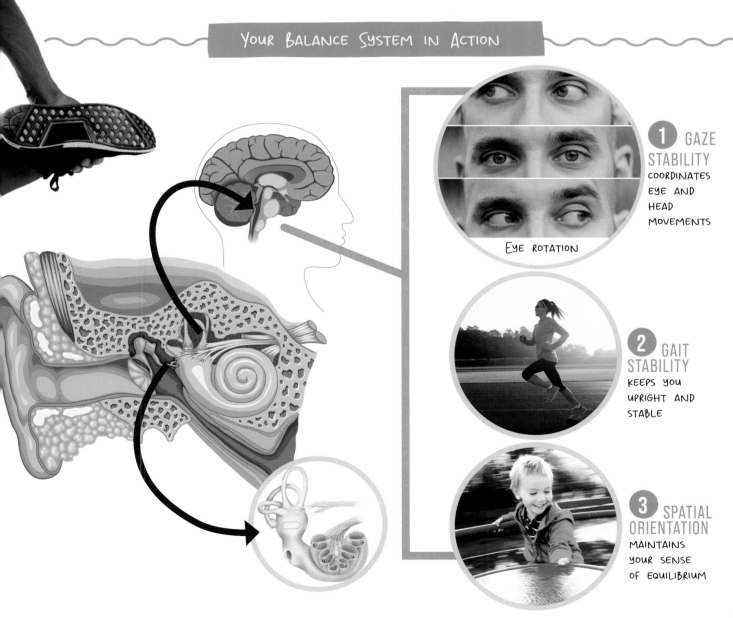

YOUR BALANCE SYSTEM IN ACTION

EYE ROTATION

1 GAZE STABILITY
COORDINATES EYE AND HEAD MOVEMENTS

2 GAIT STABILITY
KEEPS YOU UPRIGHT AND STABLE

3 SPATIAL ORIENTATION
MAINTAINS YOUR SENSE OF EQUILIBRIUM

38
How does the body prepare you for sleep with changes in light?

Instinct Guides When to Sleep

"Humans, like all creatures and even plants, have an internal biological clock attuned to the cycle of night and day. Our CIRCADIAN RHYTHM…."[228] This clock is located in the brain's hypothalamus. It is important for telling our bodies how much melatonin to produce; melatonin is used by our bodies to make us sleepy. Scientists have found that two aspects of light are used by this clock (instinct) to determine how much melatonin to make. One is the amount and duration of blue light entering our eyes. The other is the intensity and duration of all the light entering our eyes.

Researchers conducted various studies changing the intensity of blue light and of all wavelengths of light. They found that the amount of melatonin production was clearly related to both. Blue light and intense sunlight during the day decreased the production of melatonin and resulted in less sleepiness.

These scientists have concluded that this clock is based on what humans for millennia have experienced. A normal day for most people before the Industrial Revolution (so before about 1800) consisted of getting up at sunrise and going outside and seeing much bright light containing the blue wavelength (that's why the daytime sky is blue). When the sun is low on the horizon or only reflecting off the moon (nighttime), there is almost no blue wavelength light. So our circadian rhythm is based on the idea that blue and bright light means daytime while no blue and low intensity light means it's nighttime.

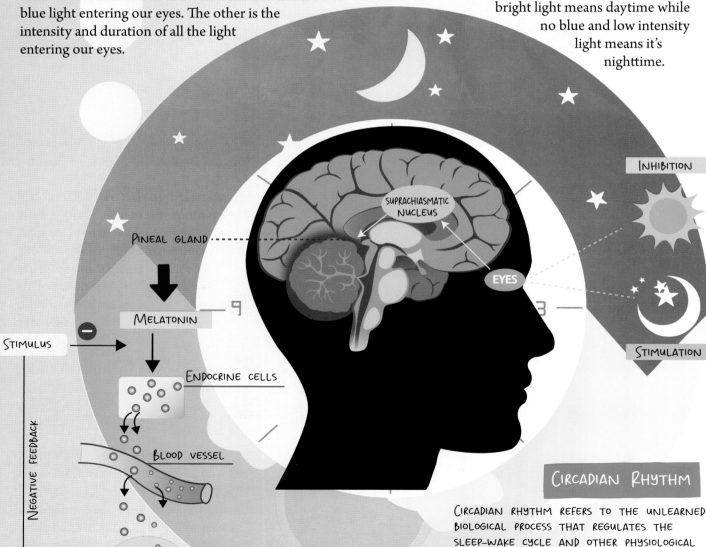

INHIBITION

SUPRACHIASMATIC NUCLEUS

PINEAL GLAND

EYES

STIMULUS

MELATONIN

ENDOCRINE CELLS

BLOOD VESSEL

STIMULATION

NEGATIVE FEEDBACK

RESPOND

TARGET CELLS

CIRCADIAN RHYTHM

CIRCADIAN RHYTHM REFERS TO THE UNLEARNED BIOLOGICAL PROCESS THAT REGULATES THE SLEEP-WAKE CYCLE AND OTHER PHYSIOLOGICAL FUNCTIONS OVER A ROUGHLY 24-HOUR PERIOD, INFLUENCED BY EXTERNAL CUES SUCH AS LIGHT AND DARKNESS.

Computer manufacturers have realized that their monitors emit a significant amount of blue light, which tends to make late night use of the computer lead to difficulty falling asleep. Therefore, some computers will have a "night mode" which reduces the amount of blue light coming from the monitors. Studies have shown that a significant improvement in sleep quality will be found by also lowering the monitor brightness.[229]

The instinct God gave us knows that melatonin is great for making us sleepy (pharmacies sell it as a sleep aid). It knows how to make melatonin and get it to the part of the body where it can make us sleepy. It also knows that blue light and intense full-spectrum light (which is sunlight) means to stop making melatonin. It also knows that dim light with essentially no blue light means to make melatonin.

As usual, this "behavior" is not really simple and would be difficult to improve. God has created us to find strength through rest, and part of this is directed through instinctive behavior that is complex, as well as beautiful.

? Why should you use the "night mode" on your computer?

Pleasure or Pain: Occasionally the same sensors that send signals interpreted as pain will send signals interpreted as pleasure! This is because God has made the instinct's programming sophisticated enough to differentiate. For example, when someone strokes your hair, you can feel pleasure or annoyance. If they stroke too slow it could be repulsive, too fast it could be annoying, but in between it could be soothing.

In this case, the instinct is listening to the special nerve endings wrapped around the base of the hair follicles. These nerve endings detect the bending of the hairs. Scientists have measured the number of signals sent out during hair stroking. They found that more signals were sent out during the stroking that felt the best. The instinct's programming was designed to consider large numbers of signals to "feel" the best. One of the researchers wrote that humans have a **"weird, complex and often counterintuitive system"** of touch circuits involved with the skin, nerves, and brain which create both pleasure and pain[230] (emphasis added).

39

In what ways does the body heal damaged tissues, muscles, and skin?

It wasn't until the 1800s that Pasteur and Lister discovered that microscopic creatures even existed. President James A. Garfield died in 1881, because American doctors did not believe Lister and kept putting their dirty fingers into the bullet hole in Garfield's back. He died of infection, not the bullet.

Not only did God know about the danger to our bodies from microbes, but He created our complex immune system to fight them. The human immune system is designed to fight off invasions by foreign matter, living or dead, and return the body to a stable, protected condition. This book will discuss just some of the amazing steps taken by the instincts controlling our immune system to protect us.

Fever: You probably have had at least one fever. You may have thought it was bad and your body needed to cool off as quickly as possible. That would be your cerebrum not being as smart as your unconscious mind. "Fevers need to be treated only if the child is uncomfortable.... There is no harm in not treating a fever."[231]

God designed us to get fevers when we are invaded by things like bacteria and viruses. Our instincts know that if our body is just a little hotter, the invisible microbes will slow down their multiplication and possibly die. "This buys more time for the immune cells to find and eliminate the invaders."[232]

When our body is invaded, special sensors detect molecules that are released into the blood from damaged cells and from the microbes. The sensors alert the brain to increase body temperature, because they detected the molecules (called pyrogens because they "generate fire" or fever).

Pyrogens are substances that can induce fever in organisms. The term "pyrogen" originates from the Greek words "pyro" (meaning fire) and "gen" (meaning producer). Pyrogens can be endogenous (produced within the body) or exogenous (derived from external sources). Endogenous pyrogens are typically cytokines, and are released by immune cells, such as macrophages and monocytes, in response to infection or inflammation.

> **?** When sick, why should you not try to reduce a mild fever?

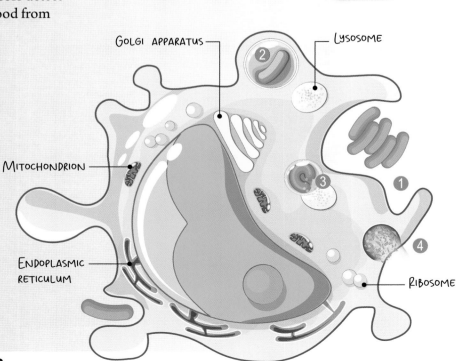

GOLGI APPARATUS
LYSOSOME
MITOCHONDRION
ENDOPLASMIC RETICULUM
RIBOSOME

MACROPHAGE

(1) MOVING TOWARD BACTERIA

(2) VESICLE FORMED AROUND BACTERIA

(3) LYSOSOME FUSING WITH A VESICLE

(4) UNDIGESTED REMAINS OF BACTERIA

Swelling

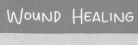
BLEEDING

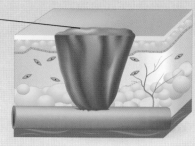

BLOOD CLOT

INFLAMMATORY

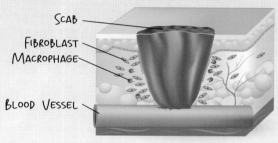

SCAB

FIBROBLAST
MACROPHAGE

BLOOD VESSEL

PROLIFERATIVE

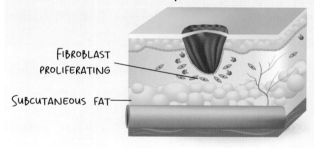

FIBROBLAST
PROLIFERATING

SUBCUTANEOUS FAT

REMODELING

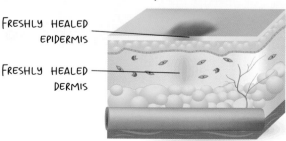

FRESHLY HEALED
EPIDERMIS

FRESHLY HEALED
DERMIS

Increased Blood Flow: The site of an injury or infection will swell, turn reddish, and feel tender and warm. This is because the instincts are sending more blood to the area. We call it an inflammation.

As soon as an injury or infection is detected, the many kinds of immune cells go to work. The ones called mast cells release chemicals that enlarge (dilate) the blood vessels near the problem. This increases the blood flow in that region. The blood will bring more immune cells, such as macrophages, to quickly attack foreign material, living or dead. Everything needed to repair any damage, including oxygen and sugar for energy, will be brought by the increased blood flow. The blood will also wash away the damaged and foreign material.[233]

Wound Healing: Wound healing is the process of repairing damaged tissue. Most of this work is done by immune cells called macrophages. The instincts send many macrophages to the damage site. They quickly start removing damaged tissue cells by eating them.

If muscles have been damaged, the macrophages release a protein that causes muscle cells to regrow.

If skin has been damaged, the macrophages release chemicals that cause new blood vessels to form. The new blood vessels will bring materials for creating new skin cells. They will also take away waste. Until the break in the skin is closed, the body is in danger of infection.[234]

40
What chemical processes work to stop infections in the body?

Infection and Memory Cells: The programming of the instincts for immunity becomes very interesting when a foreign microbe enters (infects) the body. When the body detects a foreign protein (microbes are made of proteins), the white blood cells (macrophages, B lymphocytes, and T lymphocytes) are sent to deal with it. The T cells search and find infected cells and kill them. The macrophages swallow and digest dead or dying cells and microbes (viruses, bacteria, etc.). The B cells find pieces of foreign matter left over by the macrophages and generate antibodies that will be able to attach themselves to those kinds of proteins. These new antibodies will go out, latch onto the proteins on the surfaces of viruses, and thus mark them for T cells to attack and kill.

One scientist says the "T and B cells **train themselves** to fight against this invader"[235] (emphasis added). He probably knows the details but seems to consider them so complex that he compares just the T and B cells to humans! The whole system is amazing. Here are some more of the details of the immune response to just viruses.

THE IMMUNE SYSTEM RESPONDS TO ALMOST EVERYTHING THAT GETS INSIDE YOUR BODY.

The British Society for Immunology produced a detailed description of three of the main responses of human immune systems: cytotoxic cells, interferon, and antibodies. Here are condensed versions of these three complicated methods.[236]

By Cytoxic Cells: Apparently all human cells use a system to let killer T cells know if they've been infected. When a microbe gets inside a cell, a piece of protein from the microbe will stick out of the human cell. This will be detected by the killer T cell and the human cell and microbe will be killed by special chemicals ("cytotoxic factors").

LYMPHOCYTES

B CELL
HUMORAL IMMUNITY

T CELL
ADAPTIVE IMMUNE RESPONE

ANTIBODY

CD4+   CD8+

PLASMA CELL   T-HELPER   T-KILLER

	Chemicals Released Through the Immune System		
1.	Interferons: Proteins released by infected cells to inhibit viral replication and activate immune responses against viruses.	3.	Antibodies: Proteins produced by immune cells that recognize and bind to specific foreign substances marking them for destruction.
2.	Cytokines: Small proteins that serve as messengers between immune cells, regulating inflammation and immunity.	4.	Histamine: A chemical released by immune cells in response to allergic reactions or inflammation.
5.	Chemokine: A type of cytokine that specifically help guide immune cells to the site of infection or inflammation.		

By Interferon: A second method used by the human body to fight infection is with a chemical called interferon. The moment a microbe enters a human cell, interferon is released. This chemical does two things: 1) it makes it difficult for a virus to replicate and 2) it warns other cells. The other cells will change their surfaces and thus signal to nearby killer T cells that there is a problem.

By Antibodies: When a virus is detected the instincts will attack it with antibodies. They are produced by B lymphocytes and stick to the proteins on the viruses.

Antibodies stuck on a virus will do the following:
1. Keep the virus from entering a cell.
2. Antibodies attached to viruses will stick to each other creating a bigger target for killer T cells.
3. Antibodies attached to viruses will stick to macrophages, which will then eat the viruses.
4. Antibodies will produce chemicals that will make the viruses more likely to be eaten by macrophages.

Once they know how to destroy this foreign protein, they usually clone themselves in a remarkable way. They become two pairs of expert fighters against that protein. However, one pair goes into action immediately, while the other pair (one T and one B) becomes inactive and is stored away. They are called memory cells.

The purpose of the memory cells is for the body to be ready for a rapid response against that foreign protein, if it ever invades again. Thus, future invaders will not have time to multiply and cause damage to the body.[237]

Vaccines take advantage of our instincts creating memory cells for fighting future infections. For at least 70 years, most flu vaccines have put either dead viruses, live but weak viruses, or surface material from viruses into our bodies.[238] This causes our instincts to create the killer T and B cells, which also create those memory cells that are stored for a rapid response if the live virus ever attacks.

Who does great things, and unsearchable,
Marvelous things without number. (Job 5 : 9)

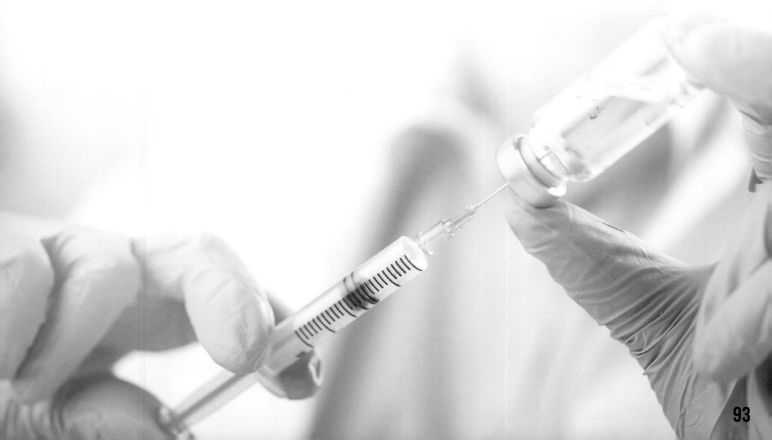

41

How important is the liver to the kidney, brain, and immune system?

"It would be difficult to overestimate the importance of the liver to the healthy functioning of the human body. It is a **remarkable organ**. The liver acts as a **processing plant,** a battery, a filter, a **warehouse** and a **distribution centre** all in one. The number and range of functions it carries **out** is **staggering**…"[240] (emphasis added). Notice the emphasized functions of everyone's liver. Each function requires some kind of programming in order to work. That programming is the set of instincts that God put into our unconscious minds. Here is a list of just some of the actions of our livers.[241]

Hepatocytes (Latin for "liver cells") make up 70% to 80% of the mass of the liver. They do a remarkable number of important jobs. They assist in making sure we don't bleed to death, protein can be stored, food is digested well, toxins are removed safely, and drugs we take will be usable. "Thanks to these main cells, we are able to fight off disease, produce waste, transport materials throughout the body and process everything from drugs and insecticides to steroids and pollutants."[242]

Liver Endothelial Cells "do not have tight membranes," which makes them able to transport other cells around the body. For example, they carry hepatocytes and macrophages where they need to go. They also improve cellular communication.

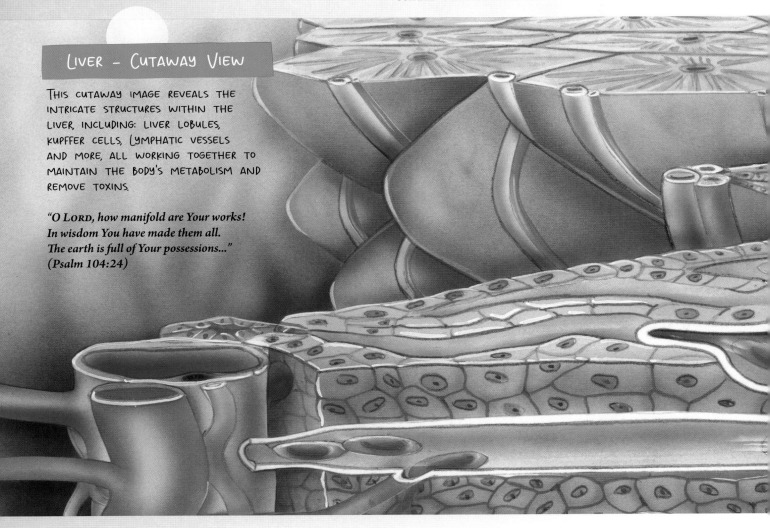

LIVER - CUTAWAY VIEW

THIS CUTAWAY IMAGE REVEALS THE INTRICATE STRUCTURES WITHIN THE LIVER, INCLUDING: LIVER LOBULES, KUPFFER CELLS, LYMPHATIC VESSELS AND MORE, ALL WORKING TOGETHER TO MAINTAIN THE BODY'S METABOLISM AND REMOVE TOXINS.

"O Lord, how manifold are Your works! In wisdom You have made them all. The earth is full of Your possessions..."
(Psalm 104:24)

Kupffer Cells in the liver hold much of the liver's lysosomes (Greek for cut or separate a body) that digest and dispose of dying cells, waste material, bacteria, and other foreign matter. "If stimulated, kupffer cells secrete mediators of the **immune response system**, and they can perform a **complex array of functions**… the kupffer cells are like bodyguards and assassins for the hepatocytes, protecting them from invaders and cell refuse" (emphasis added). Obviously, God designed this programming.

Hepatic stellate cells can be viewed as the liver's reserve army. Most of the time, these liver cells, which are only about 6% of all liver cells, do nothing except store vitamin A and a number of important receptors. However, when activated by something like an injury, they promote things like production of antibodies, killer T cells, and chemicals to reduce stress.

> "The immune system, digestive tract, kidney, brain, and cardiovascular system all depend on a healthy and well-functioning liver."[243]

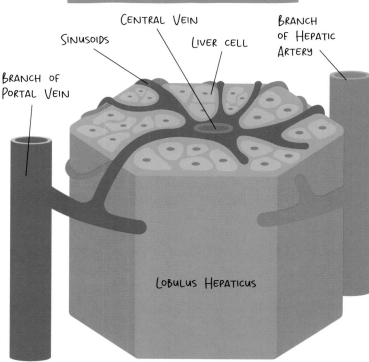

LIVER LOBULE STRUCTURE

CENTRAL VEIN

SINUSOIDS

LIVER CELL

BRANCH OF HEPATIC ARTERY

BRANCH OF PORTAL VEIN

LOBULUS HEPATICUS

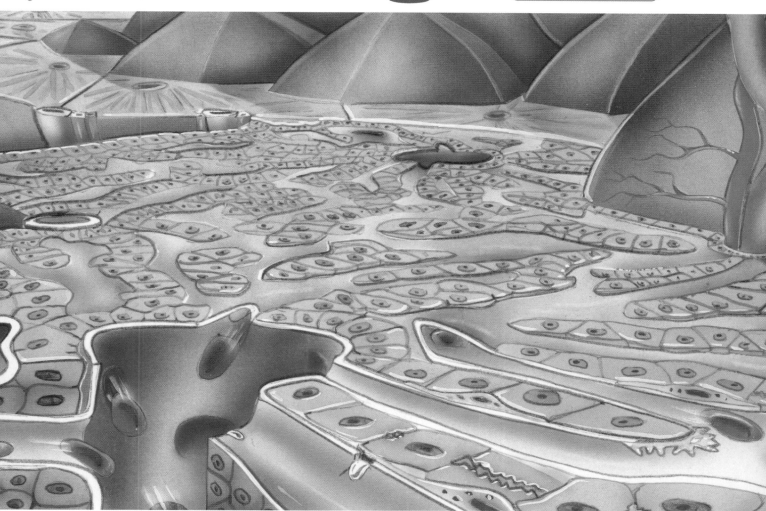

42
How do people know instinctively what to fear to stay safe?

You probably think fear is bad. That it indicates you are a weak person. That may be true of some fears, but God put certain fears into our unconscious mind in order to protect us. Those good fears are like a fire alarm: the problem is not the alarm, it's the fire. Good fears are designed to protect us. Let's look at some of them that adults have.

Fight or Flight: Scientists have found that the "vital" fight or flight reactions are based in the amygdala (a small part of the cerebrum next to the hypothalamus).[244] The "fear reaction is activated instantly — a few seconds faster than the thinking part of the brain can process or evaluate what's happening." The body is quickly preparing itself to fight or to run as fast as possible.[245] The preparation is amazing.

The amygdala activates the nervous system to immediately cause a faster heartbeat, rapid breathing for more oxygen, and an increase in blood pressure. The increased blood pressure is probably to ensure the muscles are prepared for extreme exercise. This instinct even causes you to sweat to help you stay cool when the muscles suddenly start moving.

As soon as the brain realizes there's no danger, the instinct returns everything back to normal. All of this could happen in just a few seconds (e.g., if the scare was caused by a bursting balloon).[246]

How do they know that this instinct is programmed into the amygdala (part of the unconscious brain)? Their research of human behavior has shown that if the amygdala is damaged, "humans fail to show fear or aggression."[247] As usual, this instinct is much more complicated than what has been presented.

So the Lord God said to the serpent:
"Because you have done this...
I will put enmity between you and the woman,
and between your seed and her Seed...
(Genesis 3:14-15)

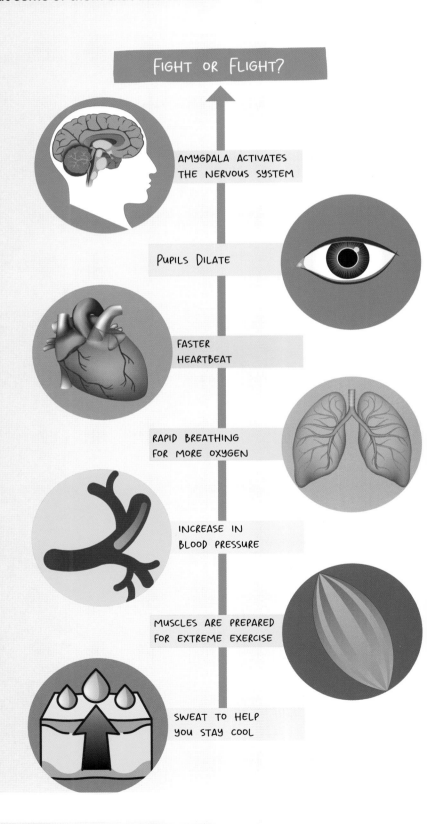

FIGHT OR FLIGHT?

AMYGDALA ACTIVATES THE NERVOUS SYSTEM

PUPILS DILATE

FASTER HEARTBEAT

RAPID BREATHING FOR MORE OXYGEN

INCREASE IN BLOOD PRESSURE

MUSCLES ARE PREPARED FOR EXTREME EXERCISE

SWEAT TO HELP YOU STAY COOL

Fear of Spiders and Snakes: This great system to protect you by rapidly determining how much and what kind of danger depends on one very important idea: exactly what is a danger to you? Stop and think about it. What is a danger to a human is not necessarily a danger to other creatures. How do humans and other creatures know what to fear and what to try to eat? In fact, how do each of God's many, many creatures not only know quickly what to fear, but they even know the degree of fear they should have? Their built-in programming is able to determine the amount of fear they should have by things like distance, size, direction, and, amazingly, what species it is.

With humans, researchers have determined that built-in programming warns humans about spiders and snakes. Studies have shown that people who have never seen a snake are afraid of them. One set of studies showed adults and children pictures of creatures like frogs, flowers, and caterpillars with either spiders or snakes, but all of the same color. Both groups could detect the spiders and snakes more quickly than each of the other creatures. They concluded that "While babies and very young children do not usually fear snakes, they are unusually skilled at detecting them and show a predisposition to learn to fear snakes" and spiders.[248]

10 of the Most Common Phobias		
1	Arachnophobia	an intense fear of spiders and other arachnids
2	Ophidiophobia	an intense fear of snakes
3	Acrophobia	an intense fear of heights
4	Aerophobia	an intense fear of flying
5	Cynophobia	an intense fear of dogs
6	Astraphobia	an intense fear of thunder and lightning
7	Trypanophobia	an intense fear of injections
8	Social phobia	an intense fear of social interactions
9	Agoraphobia	an intense fear of places that are difficult to escape, sometimes involving a fear of crowded or open spaces
10	Mysophobia	an intense fear of germs, dirt, and other contaminants

43
Why do you fear not being able to breathe?

Fear of Not Being Able to Breathe: This fear covers many situations. Fear of being in a very small space (claustrophobia) or fear of drowning or fear of being smothered. God put into us, even as babies, instincts that work hard to keep us breathing fresh air. Why? Because you die within just a few minutes without it.

If you are trapped in a small, closed space, the air's concentration of oxygen will decrease and that of carbon dioxide will increase. Scientists know that an increase of carbon dioxide in the blood will increase its pH, because when carbon dioxide dissolves in an aqueous liquid (e.g., blood) carbonic acid is created. Therefore, God created this instinct with the ability to monitor the blood in the arteries coming out of the lungs. It is constantly checked by chemoreceptors for levels of oxygen and carbon dioxide and for pH. When your breathing is restricted enough, those chemoreceptors will detect it and report it to the unconscious part of your brain.[249]

There are even pH sensors inside the brain itself: acid-sensing ion-channels in the amygdala will detect pH change. All of these signals are evaluated by the unconscious mind which will trigger the instinct's many mechanisms for trying to get more air.[250]

CHEMORECEPTORS

BRAINSTEM RESPIRATORY CENTER

BREATHING MUSCLES

ALVEOLAR VENTILATION RATE

BLOOD pH P_AO_2 P_ACO_2

FEEDBACK LOOP OF RESPIRATORY CONTROL

Fear of Drowning: The fear of drowning is especially amazing when you consider that baby spends its first 9 months "underwater" in the womb. For the last 8 weeks in the womb, the baby "breathes" the amniotic fluid. From the first minute outside the womb, now disconnected from the oxygen-providing umbilical cord, the baby will die if he inhales even a tiny amount of liquid! So, God created this strong reaction to inhaling liquid, even though for 8 weeks he was doing that very act (inside the womb)!

The instinct to try to prevent drowning involves several reflexes, but it doesn't take over until the conscious mind has had a chance to get you to safety. This is when the swimmer in trouble may be waving and yelling for help. However, once the subconscious mind determines that you're about to breathe in liquid and die, the instinct takes over. The goal is to keep only air coming into the mouth.

Some of the reflexes involve the arms being outstretched and pushing down on the water while the head is tilted back, and the mouth is open trying to breathe every time it gets out of the water. The body will be straight, and the feet will appear to be climbing a ladder.[251] Once the body submerges, the instinct will make it hold its breath. As the time of holding the breath continues, the instinct will cause what some victims have described as "unbearable pain, with a feeling that their head is about to explode."[252] All of this is to try to save your life, which can end very quickly without fresh air.

Here are some common human instincts of survival when someone falls into cold water:

COLD SHOCK RESPONSE: THE IMMEDIATE RESPONSE TO SUDDEN IMMERSION IN COLD WATER IS THE COLD SHOCK RESPONSE.

COLD WATER IMMERSION REFLEX: THIS REFLEX CAUSES PERIPHERAL BLOOD VESSELS TO CONSTRICT, REDUCING BLOOD FLOW TO THE SKIN AND EXTREMITIES.

HYPERVENTILATION: THIS IS AN INSTINCTIVE ATTEMPT TO INCREASE OXYGEN INTAKE TO MEET THE BODY'S DEMANDS DURING THE INITIAL SHOCK OF COLD WATER IMMERSION.

SWIMMING OR TREADING WATER: ONCE IN THE WATER, THE INSTINCTIVE RESPONSE IS TO TRY TO STAY AFLOAT AND KEEP THE HEAD ABOVE WATER.

SURVIVAL FLOATING: THIS INVOLVES LEANING BACK, KEEPING THE FACE AND AIRWAYS OUT OF THE WATER, AND USING SLOW MOVEMENTS TO REDUCE HEAT LOSS AND INCREASE SURVIVAL TIME.

IT'S IMPORTANT TO NOTE THAT WHILE THESE INSTINCTS CAN KICK IN, THE ABILITY TO RESPOND EFFECTIVELY IN COLD WATER SITUATIONS MAY ALSO DEPEND ON FACTORS SUCH AS SWIMMING ABILITY, PRIOR TRAINING, CLOTHING WORN, AND THE DURATION OF COLD WATER EXPOSURE.

44

Why can't scientists create life?

Survival, No Matter What:

Atheists believe that many miracles happened as life started from nothing and became what we see today. They believe **nothing** created life, so scientists should be able to repeat it. However, after all these years of experimenting, scientists still cannot create life!

Many atheists say they believe life came from a primordial soup. Think about it. If **nothing** could create life from a solution of chemicals that have never been alive, why can't scientists take "simple" creatures that have just died and bring them back to life? If **nothing** was able to put life into a soup of chemicals that has never been alive, why can't the scientists return life to "simple" life that has just died? If a scientist ever brought any living creature back to life, he would win the Nobel Prize. They can't do it. But they believe **nothing** did it.

Not only do they have to believe **nothing** did what they still cannot do, but they have to believe that the first "life" also multiplied itself **before** it died. Think about it. If that first miracle lifeform did not multiply itself somehow, then atheists have to believe **nothing** created life **again**! This would need to repeat until there was multiplication of the lifeform.

As Dr. Frank Turek says, "I don't have enough faith to be an atheist."[253]

In the real world that God created and Adam and Eve corrupted, there are many ways to die. Therefore, GOD HAS PUT INTO ALL CREATURES, EVEN PLANTS, THE INSTINCT FOR SURVIVAL. "**All living things** prioritize their own survival above all else and **will do what is necessary to stay alive**"[254] (emphasis added). The fact that this is true of ALL living things is really amazing. There is such variety in the creatures that God made, yet they all have instincts to try to preserve their own lives.

It is even more remarkable when you consider humans' instinct for self-preservation. Our desire to survive, no matter what, often appears as what we call "fear." We are surprised when we hear of someone risking their life, and possibly dying, for someone else. Romans 5:6–8 summarizes it nicely:

For when we were still without strength, in due time Christ died for the ungodly. For scarcely for a righteous man will one die; yet perhaps for a good man someone would even dare to die. But God demonstrates His own love toward us, in that while we were still sinners, Christ died for us.

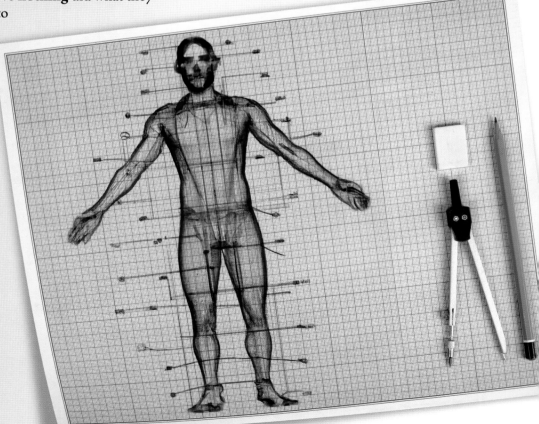

"You are worthy, O Lord,
To receive glory and honor and power;
For You created all things,
And by Your will they exist and were
created."(Revelation 4:11)

Christ's time in the Garden of Gethsemane the night before He was killed because of our selfish disobedience ("sins") shows that He had that same instinct of self-preservation that we have. The Bible tells us He sweat drops of blood (doctors say this was from stress and anxiety) as He asked the Father if this "cup" of death could be taken away. Yet, His love for all of us overcame that instinct, and He died for us. That is why the symbol of Christianity is the cross: a symbol of torture and death showing His love for us, to die in our place.

"Simple Life" Tricked Scientists:

Evolution teaches that life came from non-life and went from "simple" to "complex" over millions of years with no supervision by any Intelligence (many evolutionists will try to avoid the most difficult problem, life from non-life, by saying "evolution theory" does not deal with the origin of life). One reason Darwin was able to convince scientists in the 1800s that evolution might be true is because most thought single-celled animals were just membrane-balloons full of jelly with a nucleus and a nucleolus: "simple life."[255]

Today, scientists know better. They know that those single-celled animals, such as a paramecium, are very complex. As professor and DNA expert Robert Shapiro said in his book, *Origins,* bacteria are "among the smallest" organisms on our planet; they are much smaller than paramecia. They are so small that Darwin did not even know they existed. Yet "the interior of the bacterium shows a number of complexities, but far fewer than those present in the paramecium or human cell." Shapiro went on to describe the paramecium as having a **"bewildering array of sacs, tubes, and other structures… (and) a great richness of inner detail…"**[256] (emphasis added).

This book is designed to give the same enlightenment to readers about the "simple" behaviors called instincts.

The next section focusing on animals will provide enlightenment to readers about these "simple" behaviors we have come to call instincts.

But now ask the beasts, and they will teach you;
And the birds of the air, and they will tell you;
Or speak to the earth, and it will teach you;
And the fish of the sea will explain to you.
Who among all these does not know
That the hand of the Lord has done this,
In whose hand is the life of every living thing,
And the breath of all mankind? (Job 12:7–10)

SECTION 2:

ANIMAL

45

How do creatures know what atmosphere to look for and what to avoid?

As Genesis chapter 1 says, God created all of the animals and then created humans to "have dominion" over them. He wants humans to not only live with the animals, but to control and help them to live healthy lives.

One of the ways God has done His part in helping humans live a healthy life is by giving us the many instincts we have; some of which we just examined. He has done the same thing with the animals. He has taken the huge variety of animals and created many instincts in all of them. As we just saw in humans, most of these instincts in animals are vital for their survival. As with the human instincts, you may have known of some of these behaviors but never realized how complicated and important they are. So, let's examine just a few of the animal instincts.

Land animals need air to breathe, or they die. Sea creatures need water to breathe, or they die. However, almost all of the sea creatures would die if they were in rivers. This is because they need the large amount of salt that is in sea water. Almost all freshwater creatures would die if they were in the ocean, because of the large amount of salt.

Apparently, God wanted creatures to be in almost every place on the surface of this planet. As biologists have searched for living creatures, they have found life almost everywhere. News articles often report on the surprise to a scientist of the discovery of life in a very extreme environment. To find life in an extreme environment means that that kind of creature has the complex set of instincts (programming) necessary to survive and reproduce in such an environment.

Every creature must know what atmosphere to look for and what to avoid. Then it must know how to "breathe" that atmosphere. Each creature has a specific way it gets what its body needs from its atmosphere. It also will be able to detect whether the atmosphere is acceptable. If it is not acceptable, the creature will have programming to get its body into the correct atmosphere. Amazingly, God has put this crucial information into every creature. Let's look at some examples.

ROCKSKIPPER FISH – A SPECIES OF AMPHIBIOUS FISH THAT CAN WALK ON LAND BY HOLDING WATER IN IT'S MOUTH TO BREATHE

When a fish accidentally gets out of the water, its instincts begin to work. Fish are not designed to breathe air. In fact, the gills that get oxygen from the water that passes through them will collapse because air is less dense than water and won't support the very flexible gills. So even though air has oxygen, the surface of the collapsed gills is not exposed to it. The fish will quickly suffocate.[257]

FISH FLOP TO GET BACK TO WATER

The instinct uses the fish's only means of movement to try to get back into the water. It will keep rapidly bending its body as if it were swimming. Usually, this is successful in moving the fish back into water.[258]

GILL ARCH
BLOOD FLOW
GILL FILAMENTS
WATER FLOW
HOW FISH BREATHE

Whales and Dolphins Breathe Air Yet Live Underwater: Whales and dolphins are mammals who live in the ocean. So they need to breathe air even though they live most of their lives under water. God has designed them so that the highest point on their backs is where they breathe in air. These BLOWHOLES are closed by a flap of skin and are always closed except while breathing.

The instincts controlling breathing have the blow holes open only when air contacts them. That is why whales and dolphins almost never inhale water. They can still drown by not being able to get to the surface of the water fast enough. This can happen, for example, when a dolphin is trapped in a fishing net.

When a baby whale is born, under water, the mother whale will get under it and push it up to the surface of the water. The instinct in the baby will then make it breathe in the air.

Because whales and dolphins need to spend much time under water, God has made their bodies better than those of humans for holding their breath. For instance, their lungs are proportionally bigger than those of humans, so each breath moves more air. Their blood cells can hold more oxygen. They have a higher tolerance for carbon dioxide in the blood. **When they are diving, the instinct will have the blood,** and precious oxygen, **only go to** the "parts of the body that need oxygen — the heart, the brain and the swimming muscles.** Digestion and any other processes have to wait"[259] (emphasis added).

HOW DOLPHINS BREATHE

THE TRACHEA DIVIDES INTO TWO BRONCHI, ONE FOR EACH LUNG.

AIR MOVING IN AND OUT OF THE ANIMAL PASSES THROUGH THE BLOWHOLE. THE MOUTH IS NOT CONNECTED TO THE TRACHEA

THE HEART PUMPS BLOOD TO THE LUNGS, AND THEN AROUND THE REST OF THE BODY.

THE LUNGS ARE THE SITE OF GAS EXCHANGE. OXYGEN MOVES INTO THE BLOOD AND CARBON DIOXIDE MOVES IN THE OPPOSITE DIRECTION.

46

Why must every creature have unique instincts for finding and eating its food?

The original plan of God was for all humans and animals to be vegetarians (Genesis 1:29–30).

FOOD CHAIN

When Adam and Eve disobeyed God's explicit commands, one of the consequences was everything living was now going to eventually die. Another consequence was that life was going to be hard for the humans (Genesis 3:16–19). Apparently, this human disobedience also resulted in animals and plants killing each other and violent "natural" events, because God tells us in the Book of Romans that *the creation itself also will be delivered from the bondage of corruption… For we know that the whole creation groans… even we ourselves groan within ourselves, eagerly waiting for the adoption, the redemption of our body."* (Romans 8:21–23)

Even though man's sin brought suffering, God still directed the world as He willed it. In this way the ecosystem was formed. For example, He took the killing by the animals and plants and made it useful. Essentially always, plants and animals kill only for food and/or protection. The killing for food is what we will now examine.

The thousands of different kinds of creatures on this planet have a wide range of food sources, which together make the earth's ecosystems work beautifully. Almost everything living on the earth's surface can be food for something else; thus, when anything dies there will be some other creature ready to consume the dead body and use it for something else. This is another marvelous aspect of God's creative ability.

Therefore, each creature has its own special diet necessary for survival. Amazingly, physically, each creature has everything it needs to be able to find, catch, and turn its food into energy that keeps it active. However, without the proper instincts it would die! Each creature has special sets of instructions (instincts) for finding, catching, and eating its food.

ECOSYSTEM (EE-KOH-SIS-TUHM) NOUN: LOCATION WHERE THE RELATIONSHIPS OF ORGANISMS WITH EACH OTHER AND THEIR ENVIRONMENT TAKES PLACE.

This process of obtaining energy and body building materials (food) is even more complex. First, each creature must have programming (instincts) that lets it know it needs to find food.

HUNTING

SMELLING

As each creature searches for food, it must avoid becoming food for some other creature, so it has a set of specially designed instincts to protect it from other creatures (predators). Also, it must be able to avoid eating incorrect things that could kill it; thus, God added instincts that will detect and reject things that would hurt the creature if taken inside it.

Think about how amazing this all is. Electrical devices like your cell phone or mechanical equipment like an automobile do not know what to do when they are running out of energy. Some very sophisticated equipment will have programming to do a complex process of recharging itself. However, God has made almost every creature able to not only detect needs, but to actively take care of all its needs!

VULTURES HAVE STRONG STOMACH ACIDS THAT ALLOW THEM TO CONSUME DECAYING FLESH THAT WOULD BE HARMFUL OR EVEN DEADLY TO OTHER ANIMALS. THEIR UNIQUE ADAPTATIONS HELP THEM AVOID CONSUMING FOOD THAT IS TOO FAR GONE IN THE DECOMPOSITION PROCESS, REDUCING THE RISK OF INGESTING HARMFUL BACTERIA AND TOXINS.

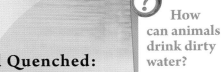

47

What are the two kinds of signals to the brain that make creatures "hungry" ?

? How can animals drink dirty water?

Thirst Sensed and Quenched:
The ability of all animals to sense their internal need for water and to generate specific behavioral responses to fulfill the need is called by a researcher **"most remarkable … this phenomenon."** He is amazed that the regulation of fluid balance can be triggered by as little as a one percent increase in salinity (salt concentration). "Small changes in the composition of the blood become a potent and specific motivational drive" organized by the instinct to find water[260] (emphasis added).

The method of detecting the need for water is with a **"specialized part** of the animal brain" that collects and monitors information such as "whether the animals have fed or drunk recently, changes in body water and salt levels, blood volume, etc." As needed, that part of the brain motivates the animal to seek water[261] (emphasis added).

The quenching of thirst in animals can be understood as being in two phases. The subconscious brain is monitoring many actions of the animal. In the first phase, some sensors determine when the animal should stop drinking water. In the second phase, other sensors determine whether more water is actually needed. The initial sensors, which stop the drinking, sense "inputs in the mouth, pharynx, esophagus, and upper gastrointestinal tract" and estimate when enough water has probably been ingested. Once the ingested water has gotten into the blood stream, other sensors measure "osmolality, fluid volume, and sodium balance" to confirm to the brain that fluid balance has occurred.[262]

OSMOLALITY: A MEASUREMENT OF THE CONCENTRATION OF PARTICLES FOUND IN BLOOD, URINE, OR STOOL.

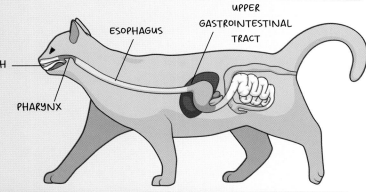

Have you ever wondered how animals can quench their thirst by drinking "dirty" water?	
1	God has put into most animals a "finely developed sense of smell" and the instinct to know which smells indicate what should not be ingested.
2	Most water sources outdoors are not as contaminated as we tend to think.
3	You have probably heard that overseas travelers should not drink the local water. Realize that the people who have lived in those places all their lives have no problem with the "local" water. Their bodies' immune systems have adjusted to the typical microbes that live in their water. The same is true for wild animals and their sources of water.
4	There's also the possibility that stomach juices are more acidic in animals than in humans, so fewer microbes can survive in animals.[269]

All creatures need water, and, as described, God has provided the instincts necessary to fulfill that need. To further amaze yourself, look at the amazing variety of ways that the many differently shaped creatures are easily able to get water into their bodies. For example, elephants use their trunks, cats bend their tongues backwards (compared to humans), bats use sonar to find it and scoop it up as they fly by, dogs splash it all over, and birds fill their bills and tilt their heads back for it to run down their throats.[263] Giraffes' long necks and height required God to put "a **complex system**" of one-way valves in their neck blood vessels to prevent too much blood pressure on the brain while drinking, and those veins are thicker and stiffer to prevent leaks and to act like a human's space suit to "aid blood flow in shifting gravity"[264] (emphasis added).

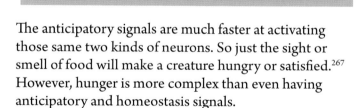

The hunger instinct in animals is amazing: How does a creature know it needs food? Scientists have recently learned that it is more complicated than "simply" monitoring things like blood (fluid) sugar level. They learned that there are two kinds of signals to the brain that will make the creature "hungry." Things like sugar availability are what has been discussed earlier: homeostasis. However, researchers found that there are signals they call anticipatory: sights, smells, and memories of food.

The homeostasis signals "fluctuate slowly over minutes to hours in accordance with changes in nutritional state."[265] These signals are from hormones, mainly leptin and ghrelin. Generally, leptin, which is made in fat cells, will make a certain kind of neuron tell the brain the body has enough energy stored. Ghrelin, made in the stomach, will make a different kind of neuron tell the brain the body needs more energy.[266]

The anticipatory signals are much faster at activating those same two kinds of neurons. So just the sight or smell of food will make a creature hungry or satisfied.[267] However, hunger is more complex than even having anticipatory and homeostasis signals.

Most creatures are strongly attracted to sugars, amino acids (which are used to create protein), and salts. Researchers found that if mice are fed a diet missing one or more of the several necessary amino acids, the mice will "reduce feeding within 20 minutes." The researchers think it is to prevent the mice bodies from scavenging stored protein for muscle building. They said, **"surprisingly"** these mice will quickly develop an appetite for the specific amino acids that are missing[268] (emphasis added).

God has so intricately crafted the minds and bodies of His creatures!

48

How can certain animals find their food using sonar?

Soon after animals are born, they must find their own food. Otherwise, they will starve and die. The instincts that guide each of them are remarkable when you remember that each kind of creature has a different menu. Not only has God made each kind of creature have the necessary body parts and digestive system for eating the food they must eat to survive, but God has given them the instincts that guide them on how to find that food, how to catch it, and how to make it part of their bodies.

How do animals find food sources by odors?

The instincts for finding food would be the programming for using the various body parts God designed for them. This would be things like using their sense of smell, sonar, hearing, or sight to find food. The instincts for catching food would be the built-in ability to use their body parts to do that. This would be things like using the long tongues of frogs that catch flies or the wings of hummingbirds that allow them to hover next to a flower as they suck out the nectar. The instincts for knowing how to make the food part of their bodies would be knowing how to bite, chew, and swallow. Now this book will look at just a few examples of the instincts for finding food.

Animals Using Smell to Find Food: Even though some animals have a very good sense of smell, it is difficult to determine the direction a smell is coming from. There are many factors that can affect an odor's path from its source (the potential food) and the odor sensor on the predator. Wind can stop and start and even change directions. There might be more than one source of the same odor. So, determining the direction of an odor "requires **computations by the neural system** since the receptor organ does not provide immediate directional cues…. (T)hese computations may involve the input from other sensory modalities or require **considerable integration of information** from the sensory input over extended time periods"[270] (emphasis added).

Researchers have studied lobsters because they have odor receptors on their antennae. They think they understand its mechanism, but they're not certain. However, they have a summary statement worth considering: "**most animals that rely on olfactory (smell) cues** for their orientation have developed **specific behavioral strategies** for tracking the source of an odor"[271] (emphasis added). They are acknowledging that the animals that use smell to find food have built-in programming (instincts) for this vital function.

1 LOBSTER ANTENNAE FOR TOUCH

2 AND 3 SMALLER SETS OF ANTENNAE FOR LOCATING FOOD

Animals Using Sonar to Find Food:

You've probably heard that bats use sound to find things (called sonar). Biologists have found that God has put that ability into several other of His creatures. "Bats, dolphins, whales, frogs, and various rodents use high-frequency sounds to find food…."[272] This means that each of these kinds of creatures is so proficient at finding food by using sound that biologists have been able to observe them many times.

The variety that God has put into His creation continually surprises.

Bats Use Sonar to Find Food: At night when bats are looking for food, their ears are more important than their eyes. God has given them an ability to locate things like food by using sound. They make a very high-pitched sound with their mouth, 10 to 20 pulses per second, and listen for any echoes with their ears.[273] Their brains can calculate an amazing amount of information from those returning sound waves (it's called sonar).

Earlier in this book, directional hearing for humans was described. That instinct is difficult to understand because of its complexity. This ability of bats makes human directional hearing look simple. Bats can "see" their surroundings with their ears. For example, a bat looking for insects to catch and eat can use its sonar to know the distance and even **"the size and shape of an insect and which way it is going"**[274] (emphasis added).

◎ Bat sonar waves

◉ Reflected sound waves of its prey

ECHOLOCATION: The process by which sound waves are emitted and their reflections help perceive the location of various objects.

Over a million Mexican free-tailed bats stream into the evening sky for a night of consuming insects, keeping down the population of flying bugs.

49

What "trick" do pigeons have that helps them "stablize" the world?

Many creatures use their eyesight to find food, including many types of birds. Most birds don't see well enough to recognize food from a great distance, so their instincts guide them. If they see other birds grouped together, their instinct tells them food is probably there. If they find food in a bird feeder, they will remember that. If the bird feeder becomes empty and they see a human go to it and it is then full, they will remember that a human can make it full.[275]

THE AUSTRALIAN WEDGE-TAILED EAGLE POSSESSES EXCEPTIONALLY LARGE EYES, BOTH IN TERMS OF THEIR ABSOLUTE SIZE AND WHEN COMPARED TO THE MAJORITY OF OTHER BIRD SPECIES. AS A RESULT, THIS EAGLE HOLDS THE DISTINCTION OF HAVING THE HIGHEST VISUAL ACUITY AMONG ALL KNOWN ANIMALS.

Creatures with keen eyesight typically rely on visual cues such as color, shape, movement, and depth perception to locate and track their food sources, allowing them to accurately assess and pursue prey or navigate towards available resources in their environment.

Eagles	These birds of prey have exceptional visual acuity and use their sharp eyesight to spot small prey, such as rodents or fish, from great distances.
Cheetahs	Known for their incredible speed, cheetahs have keen eyesight to spot prey across vast savannahs, enabling them to strategically plan and execute their hunting strategies.
Dolphins	These marine mammals have excellent vision both above and below the water's surface, allowing them to locate and pursue fish and other marine organisms.
Primates	Many primates, such as chimpanzees and monkeys, have well-developed color vision and acute eyesight, which aids them in foraging for fruits, leaves, insects, and other food sources in their forest habitats.

"Vision doesn't mix well with movement. All animals, from insects to eagles, have tricks to stabilize the world. Many mammals, humans included, do this with slight twitches of the eye."[276] This amazing ability in humans was discussed earlier and is called VOR (Vestibulo-Ocular Reflex). Now, one of the "tricks to stabilize the world" of pigeons and chickens will be discussed.

You might have noticed that when pigeons and chickens walk, their heads will bob up and down. Scientists have learned why they do that. It is to give the photoreceptors in their eyes enough time — about 20 milliseconds — to create a stable picture of the world around them. Humans do the stabilizing with just eye movement. Pigeons and chickens can also move their eyes, but "their longer, more flexible necks make it **more efficient** for them to do this motion tracking with their necks" (emphasis added).

This very necessary ability to not have a blurred view of the world is complicated. Therefore, God put it into all creatures almost from birth. The scientists that studied the pigeons admitted that it is an instinct that "emerged within 24 hours of hatching."[277]

(?) Why do pigeons bob their heads up and down as they walk?

50 Where did the animal programming come from regarding safe food?

How do all these different kinds of creatures know what to eat and what not to eat? For each kind of creature, there are probably several things that, if eaten, would kill it. God has put much of that information into each creature's instincts. Researchers think most creatures "have a combination of instinct, experience, and training that keeps them from consuming things that are harmful to them."[278] Let's look at just a few of them.

Appearance: Some animals will avoid certain colors or combinations of colors. For example, some creatures that eat butterflies will not eat monarch butterflies. This is probably an instinct protecting them from eating one and getting a very bad taste from the toxin monarch butterflies carry. However, they will also avoid eating good-for-them viceroy monarch butterflies whose combination of colors resemble the poisonous monarch butterflies.[279]

CARNIVORE

DO NOT EAT... BAD TASTE.

THE BRIGHT AND DISTINCTIVE ORANGE AND BLACK PATTERNS ON MONARCH BUTTERFLIES' WINGS SERVE AS A VISUAL WARNING TO POTENTIAL PREDATORS, INDICATING THEIR UNPALATABILITY AND TOXICITY. THIS PHENOMENON, KNOWN AS "APOSEMATISM," ALLOWS PREDATORS TO ASSOCIATE THE DISTINCT COLORATION WITH A NEGATIVE FEEDING EXPERIENCE AND LEARN TO AVOID PREYING ON MONARCH BUTTERFLIES IN THE FUTURE. PREDATORS EVEN AVOID VICEROY MONARCHS SIMPLY BECAUSE OF THEIR RESEMBLANCE TO MONARCHS.

QUEEN
Danaus gilippius

MONARCH
Danaus plexippus

VICEROY
Limenitis archippus

Taste: Many creatures have taste-receptor genes that guide the brain about whether something is good or bad if eaten by the creature. Animals that eat only plants ("herbivores") or only meat ("carnivores") have fewer kinds of taste-receptor genes than animals that eat both plants and animals ("omnivores"). Researchers say that "taste is an **especially important** sense for omnivorous species given that the potential range of foods, their variation in nutrient content, and the **hazards of accidental toxin ingestion** increase with the variety and complexity of the feeding strategy"[280](emphasis added). So the ability of these genes to correctly determine whether something is safe to eat or not is often a matter of life or death for all of these many different kinds of creatures with their many different kinds of foods. It is utterly amazing.

These researchers go into detail of how they believe (they imply their belief is obvious fact) that the herbivores and carnivores originally had all the kinds of taste-receptor genes the omnivores have. Over time, the herbivores and carnivores lost the genes they didn't need.[281] THE RESEARCHERS ARE NOT ALLOWED BY THE PUBLISHERS AND REVIEWERS TO EVEN POSIT THE IDEA THAT GOD GAVE EACH KIND OF CREATURE ONLY THE GENES THEY NEEDED. THEY IGNORE THE FACT THAT THEY DON'T FIND HERBIVORES OR CARNIVORES WITH GENES THEY DON'T USE. THEY ASSUME THAT MUST HAVE BEEN THE WAY IT WAS MILLIONS OF YEARS AGO, BUT THEY HAVE NO EVIDENCE OF IT.

Note that they assume that early in the evolution of these creatures, they had all of the marvelous taste-receptor genes necessary for knowing what is good or bad for all the different kinds of creatures! Where did all this important information and necessary programming come from? These researchers avoid this topic. They only discuss "natural selection" of the complex things that are already there… somehow.

OMNIVORES

HERBIVORES

THE GIRAFFE'S TONGUE IS 18 TO 22 INCHES LONG AND PREHENSILE, AND IT IS THE STRONGEST TONGUE AMONG ALL THE ANIMALS.

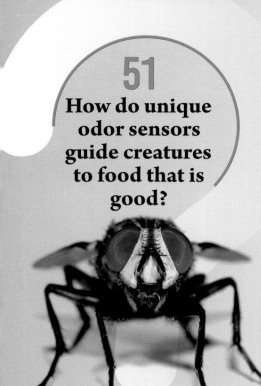

51

How do unique odor sensors guide creatures to food that is good?

Smell: "Often, an organism's **survival depends on its ability** to" use an odor to determine whether something is good for its body or may kill it[282] (emphasis added). In humans this determination is done in the amygdala of the brain. In flies it is done in a part of their brains called the lateral horn. The information built into the brains of almost all species (a few animals, like jellyfish, amazingly don't even have brains, yet they have the necessary information for survival built into them also) is what has kept their species from becoming extinct.

"They must be able to tell good from bad odors."[283] **Good odors are important for finding food. Bad smells are important for the organism to avoid rotten and toxic food.** Even tiny fly brains can distinguish between what's good and bad for their bodies. They also can distinguish the intensity of an odor, thus giving them an idea of the distance to that odor's source[284] (emphasis added).

House flies are attracted to things like feces or rotten meat, but fruit flies seek sugary, overripe fruit. They can actually smell things from over four miles away.

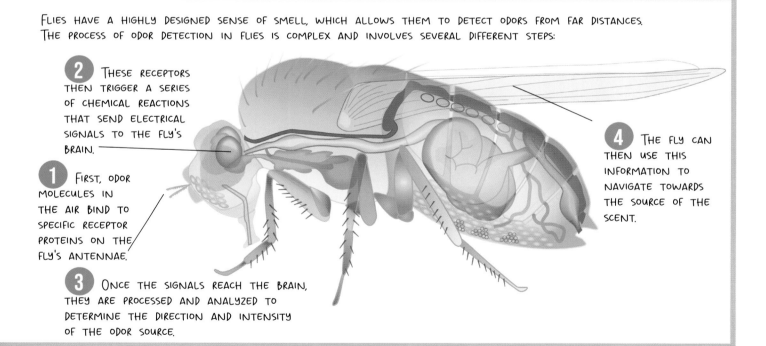

FLIES HAVE A HIGHLY DESIGNED SENSE OF SMELL, WHICH ALLOWS THEM TO DETECT ODORS FROM FAR DISTANCES. THE PROCESS OF ODOR DETECTION IN FLIES IS COMPLEX AND INVOLVES SEVERAL DIFFERENT STEPS:

2 THESE RECEPTORS THEN TRIGGER A SERIES OF CHEMICAL REACTIONS THAT SEND ELECTRICAL SIGNALS TO THE FLY'S BRAIN.

1 FIRST, ODOR MOLECULES IN THE AIR BIND TO SPECIFIC RECEPTOR PROTEINS ON THE FLY'S ANTENNAE.

3 ONCE THE SIGNALS REACH THE BRAIN, THEY ARE PROCESSED AND ANALYZED TO DETERMINE THE DIRECTION AND INTENSITY OF THE ODOR SOURCE.

4 THE FLY CAN THEN USE THIS INFORMATION TO NAVIGATE TOWARDS THE SOURCE OF THE SCENT.

Dead flies putrefy the perfumer's ointment, And cause it to give off a foul odor; So does a little folly to one respected for wisdom and honor. (Ecclesiastes 10:1)

The method that God created to accomplish this is with neurons (nerve cells) that send signals to the brain. From the signals, the brain (which has the instinct's programming) can quickly determine at least whether it is good or bad and approximately how far away it is. The instinct will guide the organism to either eat it or to get away from it.[285]

Remember that most creatures have a limited menu of what should be eaten. That is why pictures of animals in their natural habitat usually show many different kinds of animals living together peacefully. So, God had to create thousands (millions?) of different "menus" (instinctive programming) for each different kind of creature to survive. The "menus" had to be correct. They had to include all the different vitamins, minerals, protein, etc. necessary for that particular creature! Plus, the creature had to have the sensors that would detect the good and bad odors for that particular creature. This is really amazing.

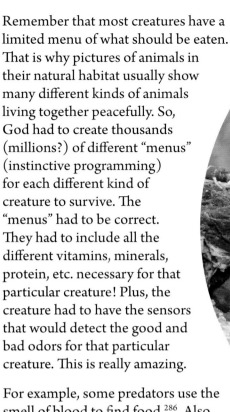

For example, some predators use the smell of blood to find food.[286] Also, animals like vultures can detect other odors from dead bodies. However, apparently, a dead body can decay too much for it to be healthy for even a vulture, so their instinct will detect that and prevent them from eating it.[287]

And God made the beast of the earth according to its kind, cattle according to its kind, and everything that creeps on the earth according to its kind. And God saw that it was good. (Genesis 1:25)

52
What instincts help brainless jellyfish hunt and kill their prey?

How do all the many different kinds of creatures know how to eat? Their body designs are so varied. Let's look at just a few of the many ways that God's creatures eat their food.

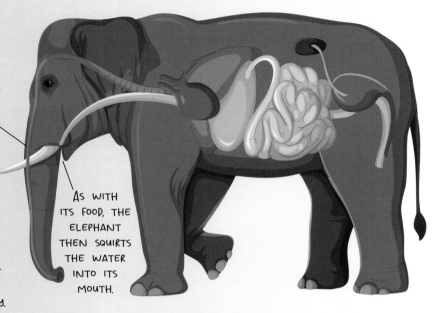

ELEPHANTS SUCK WATER UP INTO THEIR TRUNKS FROM RIVERS AND WATERING HOLES—THE TRUNK OF AN ADULT ELEPHANT CAN HOLD UP TO TEN QUARTS OF WATER!

OVER 200 PLANTS CAN BE FOOD FOR WILD ELEPHANTS, AND THESE AMAZING CREATURES EAT ON AVERAGE ABOUT 140 POUNDS OF FOOD PER DAY. THAT TAKES A LOT OF WORK!

AS WITH ITS FOOD, THE ELEPHANT THEN SQUIRTS THE WATER INTO ITS MOUTH.

Elephant Trunks: An elephant's trunk is really a nose that doubles as a hand and straw. He breathes through it, but he also uses it to pick up things. If he is hungry and food is nearby, he will use his trunk to pick it up and put it into his mouth. He will also put his trunk into a body of water and suck up the water to quench his thirst.

OTHER USES OF THE ELEPHANT TRUNK
· BREATHING
· SMELLING
· TOUCHING
· GRASPING
· SOUND PRODUCTION
· SNORKEL WHEN SWIMMING
· TAKING DUST BATH

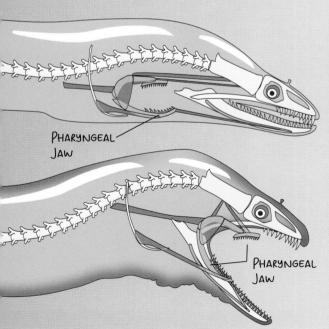

PHARYNGEAL JAW

PHARYNGEAL JAW

Moray Eels Tear Food with Teeth: Moray eels look like snakes in the water. They, like snakes, don't have arms or legs. Therefore, they are unable to grab and hold their food. God gave them very sharp, needle-like teeth that can quickly and easily tear the flesh of an animal they want to eat. This will stop the animal from moving by killing it. The moray eel can then use its sharp teeth to tear and eat the flesh.[288]

MORAY EELS POSSESS A UNIQUE ADAPTATION KNOWN AS PHARYNGEAL JAWS, WHICH ARE A SECONDARY SET OF JAWS LOCATED IN THE THROAT REGION. UNLIKE THEIR PRIMARY JAWS, WHICH ARE USED FOR CAPTURING AND HOLDING PREY, PHARYNGEAL JAWS SERVE A DIFFERENT PURPOSE. WHEN A MORAY EEL CATCHES ITS PREY WITH ITS MAIN JAWS, THE PHARYNGEAL JAWS COME INTO ACTION. THESE SPECIALIZED JAWS ARE LOCATED AT THE BACK OF THE EELS THROAT AND CAN BE RAPIDLY EXTENDED FORWARD TO GRASP AND MANIPULATE PREY. THE PHARYNGEAL JAWS HAVE SHARP TEETH AND ARE CAPABLE OF MOVING INDEPENDENTLY OF THE EEL'S PRIMARY JAWS.

Jellyfish Know How to Eat: Jellyfish are very fragile. They can be killed by a minor blow. However, God's design makes them not only able to kill fish but to eat them. The larger jellyfish have long tentacles (sometimes called oral arms) hanging down from what usually looks like a clear, very flexible bell. The tentacles can have thousands of very tiny harpoons full of several chemicals which are toxic to other animals.

When a fish swims into the tentacles, the harpoons are fired into the fish and prevent the fish from swimming away. They also inject their toxins. One toxin will stop the fish's heart, if enough is injected. Another toxin will travel to the fish's brain and cause it to stop breathing. A third toxin will get into the bloodstream and burst the red blood cells. So the probability is good that the fish will become food for the jellyfish.[289]

Once the fish has been killed or subdued, it is pulled inside the jellyfish's bell-shaped main body. The pulling is done by the muscles of the bell squeezing water in and out. Some jellyfish have short tentacles that act almost like arms and pull the fish inside. Once inside the bell, God has provided it with the necessary chemicals to digest the fish. The nutrients from the digested fish float around inside the bell, contacting and nourishing the cells that make up the jellyfish. Even this very simple creature has the programming necessary to control the bell and short tentacles to pull the trapped fish inside of its bell.

Once the fish is digested, the instinct will have the bell squeeze it out to dispose of it. This whole process of digesting and disposing of a fish is "**very fast and efficient because** it cannot eat again until the previous food has left the body"[290] (emphasis added). It makes sense that God would make the digestion faster than normal due to the limitations of this creature. However, notice that the scientist saw the need for the speed they observed, but did not conclude the obvious: there was an Intelligence behind the design of jellyfish.

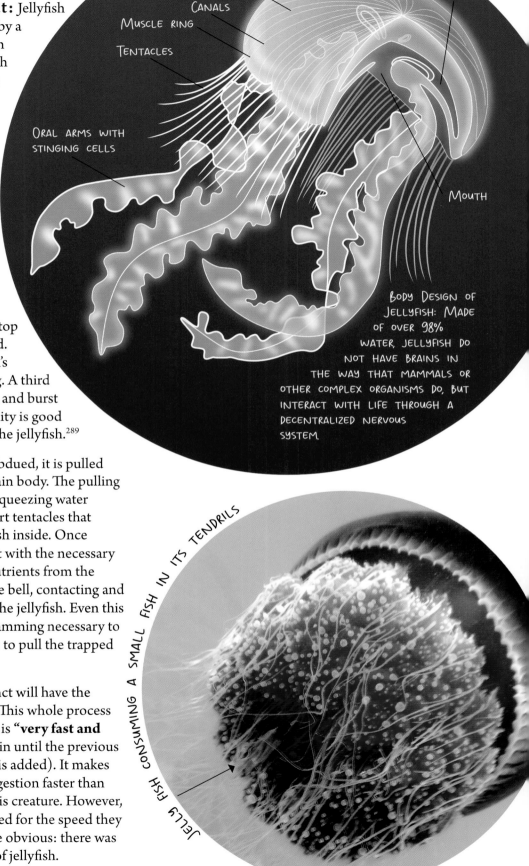

BELL

GUT

CANALS

MUSCLE RING

TENTACLES

ORAL ARMS WITH STINGING CELLS

MOUTH

BODY DESIGN OF JELLYFISH: MADE OF OVER 98% WATER, JELLYFISH DO NOT HAVE BRAINS IN THE WAY THAT MAMMALS OR OTHER COMPLEX ORGANISMS DO, BUT INTERACT WITH LIFE THROUGH A DECENTRALIZED NERVOUS SYSTEM

JELLY FISH CONSUMING A SMALL FISH IN ITS TENDRILS

53

How does the design of a cow's digestive system help it digest fiber so well?

God has made this planet one gigantic ecosystem. Every creature eats other creatures and can be food for another creature (including microbes). So, the building blocks are generally the same, carbon-based chemicals. However, He has made such diversity with those chemicals that the same digestive system cannot be used.

Different creatures need quite different nutrients. Thus, the instincts need to know what other creatures (animal and/or plant) will provide those nutrients. The instincts need to know how to gather that food and how to get it into the creature's digestive system.

The digestion of the eaten creature into the consuming creature's necessary chemicals has to be specially designed. The lining of the digesting organs needs to be strong enough to not be punctured by any sharp parts of the eaten creature. The digestive juices need to be the correct pH (acidic or basic) to break down the eaten creature into small enough molecules for use in the consuming creature, without destroying the nutrients. Once the chemicals are the useable size, the digestive system has to have the necessary method of getting those chemicals to the correct organs for conversion into body parts or stored energy for the consuming creature.

In other words, the digestive systems of all the kinds of creatures are obviously marvelously designed!

This book will only look at a few parts of digestive systems of animals.

Crab Digestive System Movement of Food: Researchers have studied how food moves through the body of a crab. Their description indicates that they see a complex program controlling the behavior, yet they give no credit to God.

They describe how the movement is subconscious, which means the crab did not learn how to do it and does not control it consciously. They say the movement of food is by "neuronal circuitry and **coordinated** firing of individual neurons…. The muscle movement pattern is a **highly synchronized, coordinated response** to food ingestion that serves to **provide the smooth movement** of foodstuffs down the digestive tract"[291] (emphasis added). This coordination of the muscles is done by the instinct God put into these creatures.

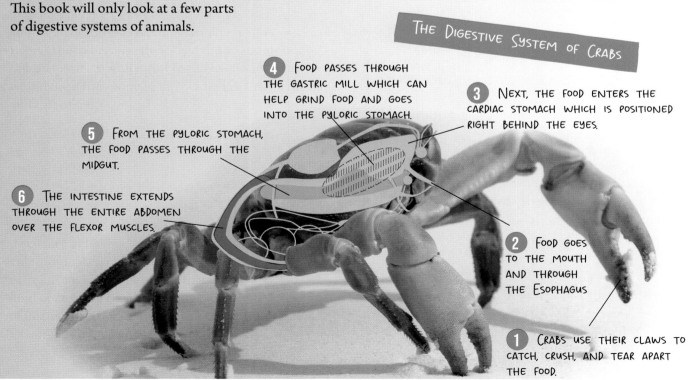

THE DIGESTIVE SYSTEM OF CRABS

4 FOOD PASSES THROUGH THE GASTRIC MILL WHICH CAN HELP GRIND FOOD AND GOES INTO THE PYLORIC STOMACH.

3 NEXT, THE FOOD ENTERS THE CARDIAC STOMACH WHICH IS POSITIONED RIGHT BEHIND THE EYES.

5 FROM THE PYLORIC STOMACH, THE FOOD PASSES THROUGH THE MIDGUT.

6 THE INTESTINE EXTENDS THROUGH THE ENTIRE ABDOMEN OVER THE FLEXOR MUSCLES.

2 FOOD GOES TO THE MOUTH AND THROUGH THE ESOPHAGUS

1 CRABS USE THEIR CLAWS TO CATCH, CRUSH, AND TEAR APART THE FOOD.

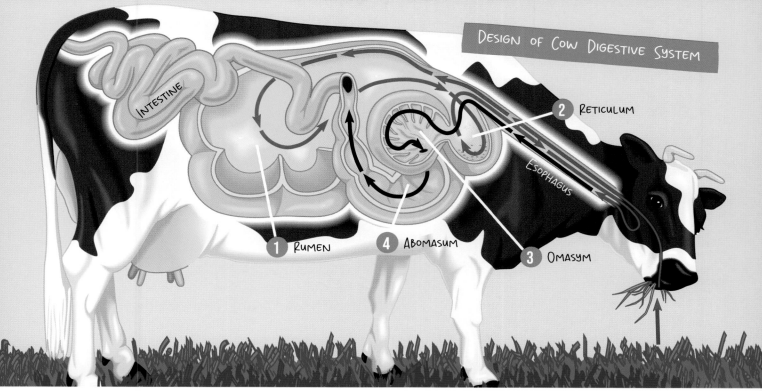

INTESTINE

2 RETICULUM

ESOPHAGUS

1 RUMEN

4 ABOMASUM

3 OMASYM

Cow Digestive System: The digestive system God put into cows (and several others such as sheep, goats, deer, elk, giraffes, and camels) is designed specifically for foods that are very difficult to digest. Basically, the cow's main food is what humans eat and call fiber or roughage. About the only value humans get out of fiber is that it scrapes clean our digestive system as it travels through us… without being digested. The cows and sheep are designed to digest it.

The method is "by their **complex digestive systems**. The way they process food, absorb nutrients and gain energy is different from other herbivores"[292] (emphasis added). The most obvious difference is that they have four stomachs instead of only one. A second major difference is that food goes down the esophagus into the first and second stomachs, and, after many hours, goes back up the esophagus for a second chewing session. Besides making the esophagus "pump" run in the reverse direction, the instinct "remembers" to make the esophagus sphincter muscle open (otherwise no food would get into the esophagus).[293]

A simplified description of the cow digestive system is as follows. The cow will bite grass and whatever is near it. It will chew it all, but just enough to swallow it. It will go into the 1 largest stomach which is full of bacteria. The larger, denser material will move to what is called the 2 second stomach though it is barely separated from the first stomach. Those two stomachs work together in churning the wet, microbe-filled mass.

During this time, the microbes are digesting the grass and other things.

At the right time, the instinct will cause the esophagus to do its peristaltic pulsing to move some of the partially digested mass in the reverse direction — back into the mouth. This mass is called a cud. So, the cow will chew its cud and then swallow this food for a second time. This time it quickly goes through the first and second stomachs into the 3 third stomach. This stomach is roughly a ball full of many folds of skin that look almost like pages of a book. This large surface area is designed to absorb much of the water and nutrients.

From here it passes on to the 4 fourth and last stomach. This one is very much like the one stomach of other mammals and humans. It has glands that secrete acid and enzymes that finish the breaking down of the various things the cow swallowed into useful chemicals that can be absorbed by the cow's intestines.

Each of these digestive system "components is **vital** in maintaining a healthy digestive process. They **must cooperate quickly and efficiently** to turn grain and plant matter into energy for the cattle"[294] (emphasis added).

Note that one of the enzymes God has the instinct secrete in the fourth stomach is not in the stomachs of one-stomach mammals. It is an enzyme that is needed to dissolve and kill the bacteria that escape from the first and second stomachs of the cow.[295]

54
Why are the instincts to sleep so widely different among animals?

? What animal sleeps the most? That would be the koala, at 20 to 22 hours per day!

Apparently, all animals and many plants have a need for sleep and the instincts that make them sleep. Scientists still don't know for certain why we all need sleep. There are many guesses, such as replenishing the brain's energy stores (glycogen),[296] helping with memory and growth and learning,[297] making repairs,[298] and improving metabolism.[299]

However, they do know that RATS COMPLETELY DEPRIVED OF SLEEP WILL DIE IN A FEW WEEKS! Insufficient sleep will make rats lose weight even though they are eating plenty of food. The rats will also get sick easier; apparently, their immune system becomes less effective.[300]

It's clear that sleep is "vital to animals' health, but its exact function and the **mechanisms that control it** are still unknown"[301] (emphasis added). The programming of these "mechanisms" in almost all creatures was put there by our Designer.

Researchers have been able to watch brain activity in various species to learn some of what goes on related to sleep. Much of the research has dealt with sleep disorders. They've learned that certain chemicals will activate neurons to tell the brain to prepare for sleep or to wake up. They've learned that there are certain genes involved with sleep and even the timing of sleep. Besides humans, scientists are studying animals such as the worm, fruit fly, and zebrafish to learn about sleep.[302] However, they still don't understand much about why all creatures need to sleep.

One Size Does Not Fit All: An animal that is asleep is more at risk of being attacked. God has designed the sleep patterns of the different kinds of creatures such that biologists can make summary statements like this:

> "As a general rule, **predatory animals can indulge,** as humans do, **in long, uninterrupted periods of sleep**…. The survival of **animals that are preyed upon,** however, depends much more critically on continued vigilance. Such species — as diverse as rabbits and giraffes — **sleep during short intervals** that usually last no more than a few minutes. Shrews, **the smallest mammals, hardly sleep at all**"[303] (emphasis added).

Remember that sleep patterns, including the length of time, are not conscious decisions of the various creatures. The above writer implies that the creatures decide how long they sleep: "can indulge." The instincts God put into creatures are designed for their particular situations in the world's ecosystem.

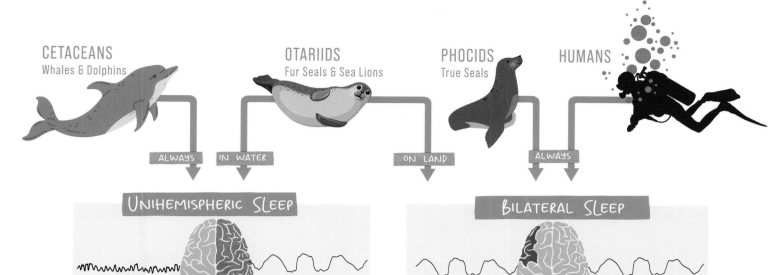

CETACEANS
Whales & Dolphins

OTARIIDS
Fur Seals & Sea Lions

PHOCIDS
True Seals

HUMANS

ALWAYS   IN WATER   ON LAND   ALWAYS

UNIHEMISPHERIC SLEEP

AWAKE
HEMISPHERE

ASLEEP
HEMISPHERE

BILATERAL SLEEP

LEFT
HEMISPHERE

RIGHT
HEMISPHERE

How Some Sea Creatures Sleep: As usual, sleep is "more complex" than has been previously described. Several different kinds of sea creatures can have half of their brain sleep while the other half is awake and alert. Scientists have attached electrodes to the brains of these sea creatures and watched the activity of the left and right halves. The electric signals clearly showed that a whole half would be doing almost nothing, while the other half was producing normal, active signals.[304] The creatures that can do this are at least white-tipped sharks,[305] whales and dolphins,[306] and seals.[307]

UNIHEMISPHERIC SLEEP is a fascinating phenomenon observed in certain animals, particularly marine mammals like dolphins. These sleep patterns differ from the typical simultaneous whole-brain sleep (called BILATERAL SLEEP) observed in most animals, as well as humans.

Unihemispheric: to sleep with only one hemisphere of its brain at a time while keeping the other hemisphere awake and alert. They sleep while still being able to swim, breathe, or surface for air.

Bilateral: a sleep pattern in which "both hemispheres of the brain are asleep at the same time."[308]

NURSE SHARK SLEEPING IN CAVE

SLEEPING SPERM WHALE DOMINICA, CARIBBEAN SEA, ATLANTIC OCEAN. SPERM WHALES HAVE A UNIQUE BEHAVIOR KNOWN AS "LOGGING," IN WHICH THEY SLEEP OR REST IN A VERTICAL POSITION, WITH THEIR BODIES NEARLY MOTIONLESS AND THEIR HEADS FACING DOWNWARD. WHY THEY EXHIBIT THIS BEHAVIORAL INSTINCT IS NOT KNOWN, THOUGHT IT MAY CONSERVE ENERGY.

55
Why would God have given horses the ability to sleep standing up?

How Some Birds Sleep: Some birds can sleep as just described about some sea creatures. They can have half of their brain sleep while the other half is awake. Birdwatchers have seen a motionless bird sitting in a tree with only one eye open. That bird almost certainly was literally "half awake." At least one reason for this is to allow the bird to quickly get away if it is threatened.[309]

Birds that are migrating will fly for days at a time. Scientists think that they probably sleep as they fly. They would be using the same method of resting half of their brain at a time. There is evidence that the Alpine Swift can fly non-stop for 200 days, sleeping while in flight.[310]

Birds' physical designs aid in their ability to fully rest while they sleep. For instance, the feet of birds that sit on tree branches are designed to wrap around and lock onto the small branches when the bird's weight pushes down on the feet. SO, A COMPLETELY RELAXED BIRD ON A SMALL BRANCH CANNOT FALL OFF! ITS TOES ARE WRAPPED AROUND AND LOCKED ONTO THE BRANCH.

You've seen the strange way that pink flamingos stand on just one long, skinny leg. Their other leg is folded up against the underside of their body. Based your own experience, you would expect that tire them very quickly. Well, flamingos sleep that way! It's hard to believe, but scientists discovered that a DEAD flamingo could be be positioned to stand on only one leg and not fall over! As the tall bird lifts one leg and leans forward, its bones just collapse "right into the position that you see when they're standing on one leg." Rotating the dead bird's body up and down on the upper leg joint could be done without the body falling over! GOD DESIGNED THE FLAMINGO BODY TO EASILY STAND AND EVEN SLEEP WHILE STANDING ON ONE LEG. The scientists pointed out that other birds, such as wood ducks and storks, also stand on one leg. Possibly this ability could be a general mechanism that many birds use.[311]

ALPINE SWIFT

Tawny Owl

Pink Flamingo

THE KOALA PROUDLY HOLDS THE TITLE OF "THE ANIMAL THAT SLEEPS THE MOST." THIS BELOVED AUSTRALIAN ICON TAKES SNOOZING TO THE NEXT LEVEL, DOZING OFF FOR 20-22 HOURS A DAY, SOLIDIFYING ITS SPOT AS THE SLEEPIEST CRITTER IN THE WHOLE ANIMAL KINGDOM.

FINCHES

MOURNING DOVE

BURROWING OWL

The programming of the sleep instinct makes birds sleep even better. On a cold night, most birds will find a place to perch. The instinct will then have it "fluff out its down feathers, turn its head around, tuck its beak into its back feathers, and pull one leg up to its belly before falling asleep." This keeps the bird quite warm, because the bare parts of the bird (the beak and one leg) are tucked into the warm feathers.[312]

Horses Sleep Standing Up:
Actually, horses only nap or doze while standing. When they need deep sleep, they lie down. God gave them an ability, similar to flamingos, to lock the major joints in their legs so that they cannot fall over. Then they can rest and even sleep while standing.

Since the horse can sleep while lying on the ground, why would God have given it the ability to sleep standing up? Scientists think that ability is needed when a horse doesn't have the protection of a human. It is difficult for a horse to get up from lying on the ground. If there is a chance of being attacked by a wild animal, it is much safer to sleep standing up.[313]

GOD DESIGNED HORSES TO BE ABLE TO KEEP THEIR EYES OPEN WHILE THEY DOZE STANDING UP, THOUGH THEIR SLEEP CYCLES ONLY LAST AROUND 15 MINUTES.

56

How are the aerodynamics of insects different from that of birds?

GOD HAS CREATED AT LEAST THREE WAYS FOR HIS CREATURES TO FLY: LIKE INSECTS, LIKE BIRDS, OR GLIDING LIKE SOME SQUIRRELS. THIS CAUSES LOGIC PROBLEMS FOR EVOLUTIONISTS. Flight is something humans wanted to accomplish for millennia but couldn't figure out how to do. While evolution says **nothing** created three different ways to fly and made creatures that are experts at each method.

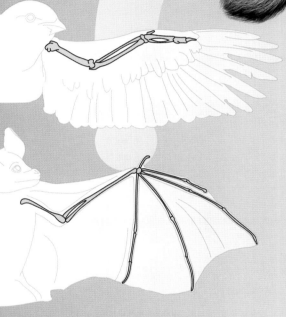

FLYING SQUIRRELS GLIDE THANKS TO A MEMBRANE BETWEEN THEIR FRONT AND BACK LEGS CALLED A PATAGIUM

At least 3 ways for creatures to fly:

1	CREATING A VORTEX – LIKE A BEE
2	CREATING LIFT LIKE A BIRD
3	GLIDING LIKE A FLYING SQUIRREL

THE BIRD SKELETON IS HIGHLY ADAPTED FOR FLIGHT. THE BONES ARE HOLLOW MAKING THEM WEIGH LESS.

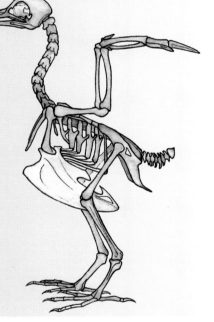

BIRDS FLAP THEIR WINGS WHICH PUSHES AIR BEHIND THEM AND DOWNWARDS.

MOVEMENT OF WINGS

IF A BIRD MOVES THE AIR WITH ENOUGH FORCE TO OVERCOME GRAVITY AND DRAG IT CAN FLY.

IT MAY NOT SEEM THAT AIR IS HARD TO PUSH THROUGH, BUT IT IS A FLUID THAT BECOMES HARDER TO MOVE THROUGH THE FASTER YOU GO.

DOWNSTROKE: THE WINGS ARE MOVED DOWN AND FORWARDS, LIFTING THE BODY IN THE AIR

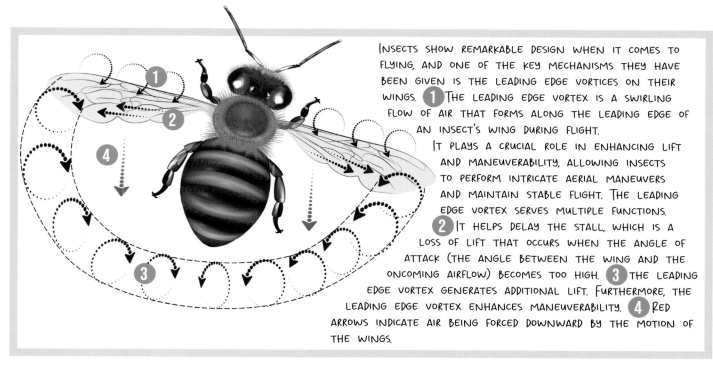

INSECTS SHOW REMARKABLE DESIGN WHEN IT COMES TO FLYING, AND ONE OF THE KEY MECHANISMS THEY HAVE BEEN GIVEN IS THE LEADING EDGE VORTICES ON THEIR WINGS. ① THE LEADING EDGE VORTEX IS A SWIRLING FLOW OF AIR THAT FORMS ALONG THE LEADING EDGE OF AN INSECT'S WING DURING FLIGHT.

IT PLAYS A CRUCIAL ROLE IN ENHANCING LIFT AND MANEUVERABILITY, ALLOWING INSECTS TO PERFORM INTRICATE AERIAL MANEUVERS AND MAINTAIN STABLE FLIGHT. THE LEADING EDGE VORTEX SERVES MULTIPLE FUNCTIONS. ② IT HELPS DELAY THE STALL, WHICH IS A LOSS OF LIFT THAT OCCURS WHEN THE ANGLE OF ATTACK (THE ANGLE BETWEEN THE WING AND THE ONCOMING AIRFLOW) BECOMES TOO HIGH. ③ THE LEADING EDGE VORTEX GENERATES ADDITIONAL LIFT. FURTHERMORE, THE LEADING EDGE VORTEX ENHANCES MANEUVERABILITY. ④ RED ARROWS INDICATE AIR BEING FORCED DOWNWARD BY THE MOTION OF THE WINGS.

The Wright brothers copied the wings of birds to create their airplane. However, insect flight aerodynamics are quite different. Instead of air flowing faster over the top of a wing than below the wing of a bird, insect wings take advantage of the physics related to turbulence around the wings.

Newly hatched insects such as locusts know how to not only fly but to maneuver as needed. This instinct is one that is similar to humans walking: it can be improved with practice. "Locusts know how to fly from birth, but they get better at it with practice, eventually learning to expend less energy to accomplish the same flight."[314]

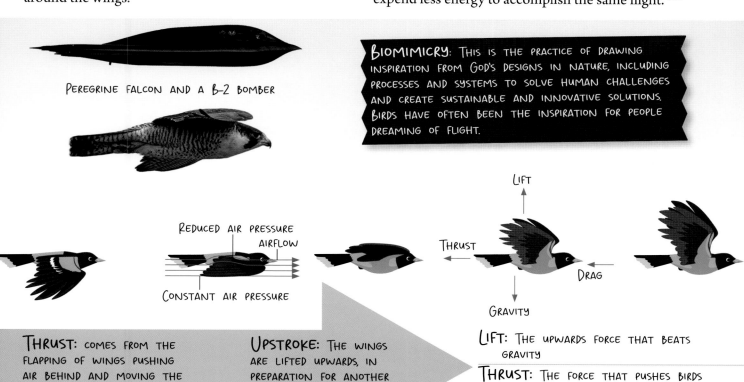

PEREGRINE FALCON AND A B-2 BOMBER

BIOMIMICRY: THIS IS THE PRACTICE OF DRAWING INSPIRATION FROM GOD'S DESIGNS IN NATURE, INCLUDING PROCESSES AND SYSTEMS TO SOLVE HUMAN CHALLENGES AND CREATE SUSTAINABLE AND INNOVATIVE SOLUTIONS. BIRDS HAVE OFTEN BEEN THE INSPIRATION FOR PEOPLE DREAMING OF FLIGHT.

REDUCED AIR PRESSURE
AIRFLOW
CONSTANT AIR PRESSURE

LIFT
THRUST
DRAG
GRAVITY

THRUST: COMES FROM THE FLAPPING OF WINGS PUSHING AIR BEHIND AND MOVING THE BIRD FORWARD.

UPSTROKE: THE WINGS ARE LIFTED UPWARDS, IN PREPARATION FOR ANOTHER DOWNSTROKE.

LIFT: THE UPWARDS FORCE THAT BEATS GRAVITY

THRUST: THE FORCE THAT PUSHES BIRDS THROUGH THE AIR

DRAG: THE RESISTANCE OF AIR PUSHING AGAINST BIRD

GRAVITY: THE FORCE THAT PULLS DOWN TO EARTH

57

How does a newborn baby kangaroo know where to go to survive?

BABY KANGAROO CRAWLS INTO POUCH

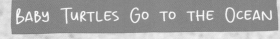

THE SURVIVAL RATE OF BABY SEA TURTLES VARIES DEPENDING ON VARIOUS FACTORS, INCLUDING THE SPECIES OF SEA TURTLE, ENVIRONMENTAL CONDITIONS, AND HUMAN IMPACTS. UNFORTUNATELY, THE SURVIVAL RATE FOR BABY SEA TURTLES IS QUITE LOW. ESTIMATES SUGGEST THAT ONLY ABOUT 1 IN 1,000 BABY SEA TURTLES SURVIVE TO ADULTHOOD.

When a baby kangaroo is born, IT WILL DIE IF IT DOESN'T QUICKLY GET INTO THE PROTECTION OF THE MOTHER'S POUCH. How does the joey know it needs to move and, significantly, where it needs to go? This is built into the joey's subconscious mind. This little creature has never walked, but it immediately can move its body with its legs. It doesn't just walk, it must climb! Once it gets to the opening of the pouch, how does it know to crawl into it?

The whole process is "simple" but amazing, especially when you remember it is crucial for survival.

A BABY WALLABY IS ONLY THE SIZE OF A JELLY BEAN

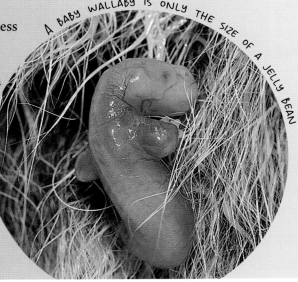

Sea turtle hatchlings provide one of the best examples of BUILT-IN PROGRAMMING NECESSARY FOR LIFE. Their eggs are buried in the sand on a seashore and abandoned by the mothers. When they hatch, they instinctively know they must start digging upward. This digging may take days! They also instinctively know not to come to the surface during daytime. The safest time is at night. When they emerge from the sand at night, they instinctively know they must rush to get into the water. Their instinct knows they are not safe until they get into the sea.[315]

Unlike human babies who take about a year to learn to walk, many animals can walk the day they are born. For instance, geese[316] and penguins[317] can immediately walk. However, for many animals, walking is a combination of instinct and conscious learning. For example, a newborn horse can walk due to the built-in programming; however, over the following weeks it is continually learning to walk better and to run.[318]

Why God gave many animals the ability to walk on the day of their birth seems obvious. Many of them are born into herds that are continually on the move. There is no time for a mother to raise a baby that can't move with the herd. THE NEWBORN MUST BE ABLE TO AT LEAST WALK ON DAY ONE, OR IT WILL BE EATEN BY A PREDATOR.

SOME ANIMALS MUST WALK ON DAY ONE

58

How do animal instincts do complicated math to determine distance?

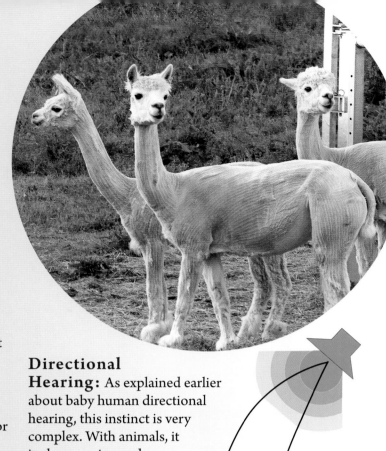

Sight for Distance to an Object: As described earlier about baby humans, sight is very important. Animals depend on it not only for protection from predators but also for finding food. If potential food or a predator is seen by an animal, the animal needs to quickly determine the distance to it. Then it must quickly determine if it is something to try to eat, or if it is a predator to run from.

Just like with humans, most animals' instincts will use parallax to determine the direction and distance to an object. The calculation needed is even more difficult for the small creatures than it is for humans, because the distance between the two eyes is so small. SOMEHOW, THE CREATURE'S BRAIN KNOWS THE DISTANCE BETWEEN THE TWO EYES AND THE ANGLE OF EACH EYE AS IT LOOKS AT THE OBJECT. THEN THE INSTINCT DOES THE COMPLICATED MATH PROBLEM to determine the distance to the object.

Directional Hearing:

As explained earlier about baby human directional hearing, this instinct is very complex. With animals, it is also amazing and more important. Animals are constantly under threat from predators and in need of finding food. Both situations rely on the animal having directional hearing.

As described for humans, the unconscious brain must do somewhat difficult math to calculate the direction to a sound. The main calculation is of the triangle whose three corners are the two ears and the sound source. With humans, the brain apparently knows the distance between the ears. The very complicated math involves the time difference between the two ears' detection of the same wave of sound from the sound source.

CHAMELEONS ARE THE ONLY LIZARDS THAT CAN SEE IN TWO DIFFERENT DIRECTIONS AT ONCE.

DIRECTIONAL HEARING IS THE ABILITY OF ANIMALS TO LOCALIZE SOUND SOURCES, WHICH HELPS PROTECT THEM BY ENABLING THEM TO IDENTIFY POTENTIAL THREATS OR LOCATE PREY WITH ACCURACY AND PRECISION.

With small animals, this calculation becomes less precise. This is because the distance between the sound sensors is much smaller. Researchers wondered how a certain kind of fly could so accurately use its hearing to locate the field crickets it lays its eggs on.

Their studies showed that the largest time difference between the two "ears" for the "songs" of the field crickets would be at most only 1.45 microseconds (humans can't detect a difference less than 10 microseconds). They found that God had modified the system for the flies by physically connecting the two membranes in their two "ears." By doing this, the time difference between the two "ears" of the same sound wave of the crickets' songs would be up to 50–60 microseconds! Thus, even the tiny fly can get directional hearing.

More research found that other fly species had slightly different anatomical versions of the connected membranes. Each species' membrane connectors would vibrate when sound waves from the particular host creature hit their sound sensors. Of course, each species' instincts are able to use these data to calculate the direction to the creature they want to lay their eggs on.

As usual, because evolution teaches that nothing designed this, the researchers assume that somehow the creatures did it themselves over a great deal of time. "Hence, through functionally convergent but **anatomically divergent evolutionary innovations,** these two fly families have **independently solved** the problem of the directional detection of low-frequency sounds…"[319] (emphasis added). What they're saying is that evolution caused tiny bones to connect hearing membranes (not just be close, but to fully connect) so that the amazing physiological effect of increasing the crucial time delay (from 1.45 to about 55 microseconds) could happen. They never mention the fact that the change in time delay would be useless if the instinct's calculation didn't take that into account!

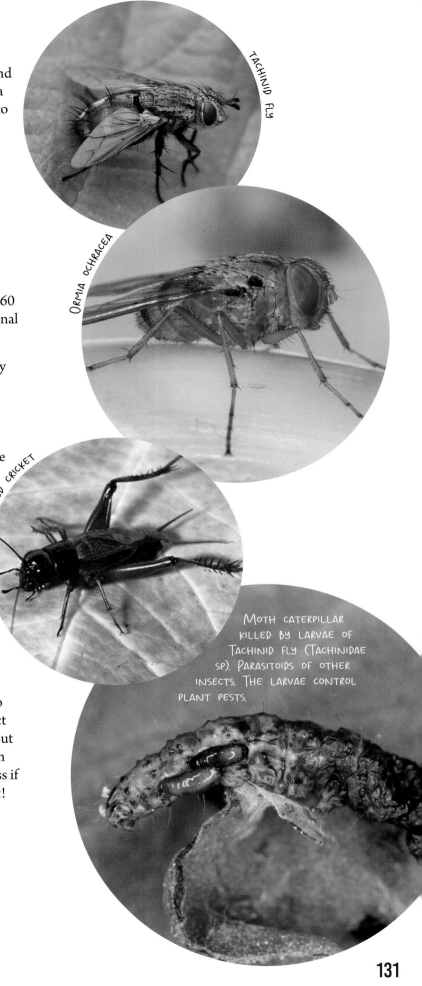

TACHINID FLY

ORMIA OCHRACEA

SOUTHEASTERN FIELD CRICKET

MOTH CATERPILLAR KILLED BY LARVAE OF TACHINID FLY (TACHINIDAE SP). PARASITOIDS OF OTHER INSECTS. THE LARVAE CONTROL PLANT PESTS.

59
How do fish in schools respond to changes in currents and group movement?

Fish Swim in Schools: You've probably noticed that some fish swim together in what humans call schools. If you watch a school of fish, you should be amazed. They act like one huge body. As the school moves, it will make sudden turns and changes in speed. Yet, there won't be any significant changes in the very small distance between the hundreds or thousands of fish in that school. This is called shoaling.

Researchers have studied this and concluded that there is a "genetic basis underlying the complex, social behavior of schooling." Two studies have confirmed this. One study showed that the fish use a structure unique to fish called a lateral line. It is a line of hair cells along each side of the fish. These hair cells are very sensitive to water currents and are thus able to detect vibrations from nearby fish. The other study compared two very similar fish species with one major difference: one species moves in schools while the other doesn't. They found a genetic difference in the two species and concluded "that schooling behavior is genetic-based and not learned." In other words, schooling is controlled by the built-in programming called an instinct.

These researchers went further in being amazed at how fish can swim together so well.

"[S]chooling requires **coordinated body positions and synchronized movement.** Fish in schools need to **sense their environment with high accuracy,** maintain awareness of their position within the school, and respond quickly to changes in both water currents and movement of the group.

Exactly how fish pull off this feat has remained elusive."[320] Apparently, only God knows!

3 remarkable aspects of fish shoaling together are:	
1	Coordination and Synchronization: Fish in shoals or schools exhibit exceptional coordination and synchronization in their movements.
2	Predator Avoidance: Schooling behavior provides significant protection against predators.
3	Hydrodynamic Advantages: By swimming in close proximity, fish create a hydrodynamic wake that reduces individual drag and saves energy expenditure.

Bird Sense of Balance: You've heard the expression, "You're running around like a chicken with its head cut off." Well, it's more true than you probably thought. Biologists have learned that most birds have a second balance-sensing organ that other animals with backbones don't have. When animals with backbones have their heads cut off, the body might twitch some, but it cannot get up and run as head-less chickens do. The difference is the second balance-sensing organ which is in the pelvis of the bird. When the head has been removed, the chicken still has the pelvic balance organ and can run around without falling down.[321]

The bird balance organ is a bulge at the bottom of the spinal cord in the pelvis. It has been concluded that this second balance organ is very helpful for birds. It helps in at least two ways. One way relates to what was mentioned earlier, a bird's bobbing of its head in order to see clearly. Because the head is bouncing around so much, it helps to have a balance organ below the neck.

A VERY STABLE DESIGN

A second way a bird is helped to have a balance organ in its pelvis relates to the fact that birds only have two feet and are usually perched on narrow branches or wires that are often randomly moving. This balance organ closer to the feet provides quicker responses to movements of the body and thus better overall balance.[322]

As usual, the researchers made little mention, if any, of the programming needed to interpret balance data from two very different locations in the bird body. This built-in control system must be very complex.

A different group of researchers found the pelvic balance organ in every one of the many different bird species they checked.[323] Since birds come in many different sizes and shapes, each species must have its own version of the instinct. Imagine the differences in those instincts' calculations necessary to keep the birds stable as they stand, walk, and fly.

? Why can chickens literally run around with their heads cut off?

60
Why would birds develop an instinct for singing?

Baby Animal Behavior Can Be Instincts Plus Training

Biologists have found that animal behavior can be a result of built-in instinct creating an action that gets refined over time and use. Several examples will be described below. For the point of this book, notice that the researchers admit that at least part of the behavior was present in the animals' genes as programming. Where would this complex information have come from? It's very hard to believe these often-crucial programming steps came about slowly over time with no Designer.

As you look at the few examples given here, think about why God would not have had the newborn animal completely proficient at a particular behavior. Why would it have needed a parent to train it? One possible reason would be to build the familial bond. This would certainly be true of humans. As you read the Book of Proverbs in the Bible, you see that God expects us to learn from the animals. This need of parental help within the Animal Kingdom would provide many examples of familial love for humans to learn from.

Bird Songs: Bird songs are beautiful for humans to enjoy, but God gave the birds the ability to "sing" for other important reasons, also. They can be used to warn each other of nearby predators.[324] They can be used to let other birds know that the singer claims an area. The different species have their own songs to help them find a suitable mate; from a distance they can locate a potential partner.[325]

Because God has given different kinds of songs to different species of birds, humans can enjoy listening to the variety and even try to recognize the species by the song.

Researchers have found that birds have an instinct for singing. However, usually the particular songs they sing must be learned by hearing them from the adult birds. "Zebra finches, for example, are **preprogrammed to learn a song,** but which song they learn depends on their early experience"[326] (emphasis added).

As usual, this is not simple. "Evidently, there is also a **strongly instinctive aspect** to what may be learned during the critical period; most birds **cannot produce every song heard** during that time, but appear to be **selective toward songs** that are produced **by other members of their species**"[327] (emphasis added).

Flying Locusts: Flying creatures will sometimes know how to fly at birth. However, they may not fly efficiently at first. That is the case with locusts. "Locusts know how to fly from birth, but they get better at it with practice, eventually learning to expend less energy to accomplish the same flight."[329]

Flying Birds: As you probably know, baby birds do not fly immediately. Their bodies are not capable of flight when they emerge from their eggs. Their wings are too small and don't have the necessary feathers; they are covered in fluffy down (good for warmth, but not flight).

As the baby bird grows, its wings lengthen and strengthen. The down is replaced by the remarkable, aerodynamic feathers. Dr. Michael Denton in his book, *Evolution: A Theory in Crisis,* said, "The feather is a magnificent adaptation for flight…. In addition to its lightness and strength the feather has also permitted the exploitation of a number of sophisticated aerodynamic principles in the design of the bird's wing."[328]

When the baby bird is developed enough, the mother bird's instincts tell her the baby bird is ready to fly. Therefore, she will usually keep encouraging the baby to walk out of the safe nest into the air… with nothing to catch him!

Eventually, the baby bird's instincts will succeed in getting him to walk out into the empty space. The instincts will quickly make the baby bird do everything necessary to fly. The first time won't be perfect. With help from the parent birds and practice, the baby will soon be an excellent flyer.

FLEDGLING STAGE: THE FLEDGLING STAGE IS THE FINAL STEP BEFORE BIRDS LEARN TO FLY. AT THIS POINT, THE YOUNG BIRD STARTS TO LEAVE THE NEST AND EXPLORE ITS SURROUNDINGS. IT BEGINS TO EXERCISE AND STRENGTHEN ITS WINGS BY FLAPPING THEM VIGOROUSLY. INITIALLY, THE FLEDGLING MAY ONLY BE ABLE TO GLIDE SHORT DISTANCES OR MAKE CLUMSY ATTEMPTS AT FLIGHT. HOWEVER, THROUGH PRACTICE AND EXPERIENCE, IT GRADUALLY DEVELOPS THE NECESSARY SKILLS AND MUSCLES TO BECOME AIRBORNE AND ACHIEVE SUSTAINED FLIGHT.

61
How can foals be born with the knowledge of how to walk?

Within the first hour after birth, foals usually attempt to stand. They instinctively push themselves up using their forelimbs, arch their back, and gradually gain stability on their feet. This process is critical for their circulation and helps them fully inflate their lungs.

Running Horses: You've probably enjoyed watching a nature video that showed the birth of a baby horse (a foal). You saw the foal struggle to stand within seconds of his birth. Usually within a few minutes the foal actually almost runs, but it is obvious that he can't do it very well.

Researchers have determined that foals are "born with the knowledge of how to walk," but "it still takes time for the foal to learn how to operate its legs."[330] So, God's instincts prepare foals by giving them the ability to walk as a newborn. Then, with the help of the mother horse and practice, the horse learns how to walk and run.

Foals can stand, walk, and trot shortly after birth. Ideally, a foal should be up and nursing within two hours of birth.

Feral donkeys and horses have been observed digging wells to access groundwater in desert environments. This behavior is known as "water-holing" or "well-digging," and it is a remarkable instinct that allows them to survive in arid regions with limited water resources. Here are some details about this behavior:

Digging Technique	Feral donkeys and horses use their front hooves and sometimes their muzzles to excavate the soil and create a hole. They repeatedly paw at the ground, removing sand, soil, and debris, gradually deepening and widening the hole. The size of the wells can vary but can be several feet deep and wide enough for the animals to access the water.
Accessing Groundwater	In desert environments, groundwater is often present at varying depths beneath the surface. By digging wells, feral donkeys and horses can tap into this groundwater, allowing them to access a relatively reliable water source. They create openings in the ground that reach the water table, enabling them to drink and satisfy their hydration needs.
Water Conservation	These wells not only provide water for the diggers themselves but also serve as a resource for other wildlife in the area. The excavated wells can accumulate and retain water, creating small oases in the arid landscape. This water can support a variety of desert-adapted organisms, including birds, reptiles, and other mammals.

62
What causes creatures to have a fear of predators built into their DNA?

Fight or Flight for humans has been described earlier. It is just as complex for animals, but probably more amazing considering the "simpler" brains and the more frequently they experience life-threatening situations. With that in mind, consider the complexity shown in this scientist's summary of FIGHT OR FLIGHT instinct in animals:

When under imminent attack, (v) **defensive systems** evoke fast reflexive indirect escape behaviors (i.e., **fight or flight**). This **cascade of responses** to threats of increasing magnitude are underwritten by an **interconnected neural architecture** that extends from **cortical and hippocampal circuits**, to **attention, action and threat systems including the amygdala, striatum, and hard-wired defensive systems** in the midbrain...[331] (emphasis added).

AS A CAT APPROACHES, A MOUSE SEEKS A WAY OF ESCAPE, A PLACE OF PROTECTION WHERE THE CAT IS UNABLE TO REACH IT.

Even in rodents, scientists have found that the unconscious mind does much assessing of the actual situation before the very rapid reaction. Studies with rodents have shown that the senses signal to various parts of the involuntary brain several important pieces of information. For example, they signal what kind of creature is detected, how far away it is, what direction it is moving, and what escape routes are available. All of the information is evaluated, and the proper part of the midbrain responds... rapidly.[332]

Let's now look at some pieces of information and actions built into the Fight or Flight instincts of various animals.

CONDYLOSTYLUS ARE A GENUS OF FLIES AND ARE THOUGHT TO HAVE THE FASTEST REFLEX RESPONSE IN THE ANIMAL KINGDOM. SCIENTISTS HAVE MEASURED ITS REFLEX RESPONSE TIME TO BE LESS THAN 5 MILLISECONDS. FOR COMPARISON, AN AVERAGE HUMAN BLINK IS ABOUT 1/3 OF A SECOND.

What's Prey and What's Predator? When you are walking in the woods and see an animal you're unfamiliar with, what do you quickly try to determine? You probably think, "Will it hurt me?" Imagine that you're a small animal and encounter another type of animal. You need to be able to decide quickly if this creature might harm you before it's too late.

It is crucial for all creatures to know what is food and what wants to make you food!

This information is put into all creatures by their Creator. If it had to be learned and somehow passed on to later generations through gene mutation, it's very hard to believe any form of life would exist today. With this in mind, I repeat what Dr. Frank Turek says, "I don't have enough faith to be an atheist!"[333]

LONG-HAIRED SPIDER MONKEY

One evolutionist, apparently desperate to find an evolutionary answer to the instinct of recognizing a predator, wrote, "We have known that many species of monkeys either have an innate fear of snakes or pick up a fear of snakes very readily. This provides a **probable mechanism.** That has been a **huge question** in the literature"[334] (emphasis added). If his mechanism is that the monkeys have to learn the fear of snakes by trial and error, he's being dishonest. Many creatures have the fears of their specific predators built into their DNA (instinct); for example, the kangaroo rat instantly performs an escape jump at the sound of a rattlesnake, "even if it has never encountered a snake before."[335]

The key phrase in the previous quote shows how little the scientific community knows about instincts. Scientists have been trying for decades to understand how evolution could inform creatures which other creatures they should fear. For evolutionists, it is "a huge question." Evolutionists are left only with questions, while those who know the Creator find the answers so clearly in Him and His great design.

RATTLESNAKE

139

63

How do animals know how to react to predators?

Scientists have been able to study the mechanisms God put into creatures to protect them from predators. "For example, many rodents have receptors in the vomeronasal organ that respond explicitly to predator stimuli **that specifically relate to that individual species** of rodent. The reception of a predatory stimulus usually creates a response of defense or fear"[341] (emphasis added). Note that the physical receptors respond to only predators "that specifically relate" to that kind of rodent. That is amazing, and obviously, not random.

God has fine-tuned the ecosystem. All of the creatures' instincts know what is predator and what is prey (food). They also know who is a friend, but that is another topic. Let's look at a few ways animals react to predators.

Animal Escape Reflexes: There are many kinds of escape reflexes that humans and animals have. When surprised by something, we and animals will react quickly. We may duck our heads (to protect our heads), jump to the side, turn and run, or possibly swing our arms at the threat. These are instinctive reflexes. They are necessary for our protection. Let's look at some reflexes God gave the animals.

C-Start in Fish and Amphibia: One of the simplest, and therefore most studied, reflexes is the C-Start in fish and amphibia. Basically, it is a study of what will cause a fish to quickly start swimming.

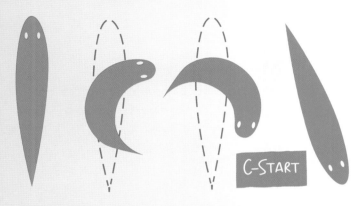

C-START

The "C-Start" name is just a way to describe the very common method that fish use to quickly escape in the opposite direction: bending their bodies into a C-shape and swimming in the new direction.

These many studies have found that most fish will react, by instinct, to a sudden change of lighting[336] or a sudden noise.[337] Researchers have determined which neurons are involved and what part of the fish brain controls the reflex.

Even this "simple" reflex is complicated enough to take into account the direction and strength of the stimulus (light, sound, etc.) and then choose a direction for the fish to swim. Some researchers have apparently succeeded in mathematically describing this "simple reflex" with a complex "set of descriptive equations written in terms of stimulus angle, magnitude and timing variables of trunk muscle contractions, and resulting escape trajectory."[338] Surely, nothing so intricate could simply be random rather than designed.

Striped marlin and sea lion hunting in sardine run bait ball in Pacific Ocean

Escape Action of Mollusks

Even mollusks have instincts that protect them from predators. The mollusk Tritonia looks like a clam without a shell; because it acts like one big muscle, it greatly resembles the last two inches of your tongue. Starfish love to eat them. The Tritonia instinct responds to protect it. Therefore, if a starfish touches a Tritonia, the instinct will rapidly react to protect it.

Biologists have studied how the instinct works. As soon as the tube feet of a starfish contact a Tritonia, it reacts with the **"highly coordinated firing of an elaborate network** of neurons in the animal's nervous system." It's a **"hard-wired** behavioral response ... (of) self-perpetuating chemical reactions"[339] (emphasis added). The observable reaction of the Tritonia could be called swimming.

The swimming results from the neurons getting in a positive feedback loop. The loop has the neurons alternating between quickly bending the Tritonia in the middle downward (pushing off the sea bottom and rising) and quickly bending it in the middle upward. This bending back and forth continues until the instinct decides the Tritonia is safe.[340]

REACTIONS AND REFLEX ACTIONS IN INSECTS ARE BOTH TYPES OF RAPID RESPONSES TO STIMULI, BUT THEY DIFFER IN TERMS OF COMPLEXITY AND UNDERLYING MECHANISMS. A PRIMARY DIFFERENCE BETWEEN THE TWO: REFLEX ACTIONS IN INSECTS ARE RELATIVELY SIMPLE AND INVOLVE A DIRECT, INSTINCTIVE RESPONSE TO A SPECIFIC STIMULUS. THEY ARE TYPICALLY MEDIATED BY THE INSECT'S CENTRAL NERVOUS SYSTEM. REACTIONS, ON THE OTHER HAND, ARE MORE COMPLEX AND CAN INVOLVE SENSORY INPUT, INTEGRATION, AND DECISION-MAKING PROCESSES. REACTIONS MAY INCLUDE MULTIPLE STEPS OR RESPONSES BASED ON THE INSECT'S INTERPRETATION OF THE STIMULUS AND THE CONTEXT IN WHICH IT OCCURS. BECAUSE OF THIS, REFLEX ACTIONS TEND TO BE FASTER THAN REACTIONS.

64
How do squid instincts help them move through jet propulsion?

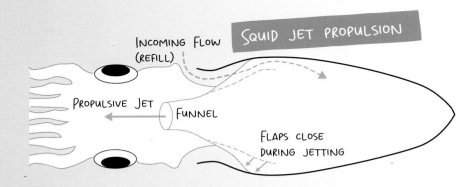

SQUID JET PROPULSION

INCOMING FLOW (REFILL)

PROPULSIVE JET

FUNNEL

FLAPS CLOSE DURING JETTING

Escape and Startle Reflexes in Earthworms: Segmented worms, such as earthworms and leeches, have instincts that protect them. Different species of segmented worms have slightly different actions available to them in their instincts. Usually, they will be able to swim or crawl to escape whatever surprised them.

The instinct will be controlled in the worm's central nervous system. It can make the worm swim by wiggling the body in an undulating manner. It can make the worm crawl by peristaltic action (as if two fingers were squeezing it near the head and started sliding down its body, still squeezing).[342]

A ROBIN TUGS AN EARTHWORM FROM THE GRASS. THE WORM, STRETCHED TO ITS LIMITS, GRASPS THE GROUND IN HOPES OF ESCAPING THE FIRM GRIP OF THE BIRD'S BEAK.

SEGMENTS

CLITELLUM

PROSTOMIUM

SEGMENTAL NERVES

BRAIN

NERVE CORD

EARTHWORM NERVOUS SYSTEM

Escape Reflex in Squid: Squid look much like giant jellyfish. However, their body and instincts provide them with a "**remarkably effective** form of locomotion through jet propulsion" (emphasis added). The main body is shaped like a bell and is mostly muscle. It can expand and contract with great force. When it expands, it pulls in a large amount of water. When it contracts, it forces the water out through a narrow funnel. This produces a powerful jet that gives the squid great speed through the water.[343]

REEF SQUID

This jet propulsion, when controlled by the squid's instinct, gives it a rapid escape reflex. When stimuli like a bright light activate the squid's axons (nerve fibers that carry signals between sensors, neurons, muscles, brains, etc.), the instinct will go into action. As usual, this is not simple. There is a "small axon motor pathway" that will quickly create a "vigorous escape jet." But there is also a "giant axon" pathway that may or may not be used. If the giant axons are used, they will be "critically timed" by the instinct to boost the force of the jet of water and thus increase the escape speed. So, the squid has "at least two escape modes in which the giant axons can contribute in different ways to the control of a **highly flexible** behavior"[344] (emphasis added).

Rapid Backwards Swim of Crayfish:

When a crayfish feels threatened, it will quickly swim backwards, away from the threat. Its instinct will cause it to sharply flex its abdomen and use its tail as a paddle. It will at the same time streamline its body by thrusting its appendages forward.

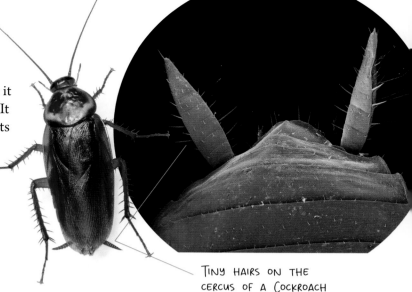

Scientists have determined that the middle and side giant nerve fibers of the crayfish's nerve cord are what cause the muscles to react so quickly. Every time these fibers are stimulated by the scientists, the crayfish will quickly try to swim backwards; this reaction will occur even if stimulated every 10 seconds.

However, a normal crayfish will not react to the same kind of threat if repeated more than a few times within several minutes. A full reaction won't occur again unless it has been "a number of hours" since the crayfish felt the quick repetition of threat. The scientists concluded that the instinct is complicated and much more useful to the animal than a simple reaction to every apparent threat.[345]

TINY HAIRS ON THE CERCUS OF A COCKROACH

Cockroach Running Away:
We have all been frustrated trying to kill a cockroach. They are often able to escape because we can't predict the direction they will run. Scientists have learned how they do that.

Built into even young cockroaches are a set of "escape trajectories" they can choose from when threatened. These escape patterns will be executed at "fixed angles from the direction of the threatening stimulus."

The choice of escape trajectory is significantly influenced by the tiny wind-sensitive hairs on the rear of the cockroach which detect air movement. These hairs connect through giant interneurons with the leg motor neurons. Thus, air direction and intensity are factors in the cockroach instinct's choice of escape trajectory.[346] Humans and predators are usually unable to guess which way a cockroach will run.

CRAYFISH FIGHTING TO GET AWAY FROM A EURASIAN COOT.

? How is a small-brained cockroach able to escape your attempts to catch it?

65

What makes an opossum playing dead release liquid that smells like a dead body?

Pufferfish Becomes Difficult to Eat: A pufferfish will puff itself up when threatened. This has the potential to frighten an attacker. It also will make a predator that wants a meal wonder if he can actually eat such a large creature.

The pufferfish's instinct knows that the fish's body is capable of enlarging significantly by just "inhaling" a large amount of water. The instinct also knows this will deter many of the pufferfish's natural enemies.[351]

Many pufferfish species also possess a powerful defensive mechanism in the form of toxins that their instincts can use. Certain organs, such as the liver and ovaries, of some pufferfish species contain a potent neurotoxin called tetrodotoxin. This toxin is highly toxic and can be lethal to predators, including humans, if ingested. The toxicity acts as a deterrent to potential predators, making pufferfish unappealing and dangerous to consume.

PUFFERFISH

PUFFERFISH MAKES HIMSELF BIG TO AVOID BEING EATEN

OPOSSUMS CAN PLAY DEAD ANYWHERE FROM A FEW MINUTES TO A FEW HOURS

Playing Dead: Several animals use a defensive strategy that includes playing dead. When they detect an animal their instincts warn them about, they will quickly pretend to be dead. Different species have slightly different versions of playing dead.

One of the creatures most famous for playing dead is the VIRGINIA OPOSSUM. In America, when a human child pretends to be asleep, the parent will often say, "Are you playing possum?" The opossum instinct can make the animal look really dead: the eyes close, the breathing appears to stop, there is no pulse, the body goes limp, the tongue hangs out, and they drool. The instinct even makes the anal glands release a liquid with the same smell as a dead body! This will often make the predator leave them alone.[347]

The EASTERN HOGNOSE SNAKE'S instinct performs a death scene like an actor in a movie. "When facing dangers, these snakes will writhe back and forth and excrete a foul odor as if they are in pain and dying. Then they will collapse on their back, slightly part their jaws, and remain still. They will stay still like this even if you poke or handle them!"[348]

The Texas indigo snake's instinct will do a simple act of being dead, but with a twist that researchers think is important. The snake will often play dead when threatened by a predator. It will coil itself up and open its mouth. The researchers think the mouth opening is an attempt by the instinct to make it harder for a predator to swallow it.[349]

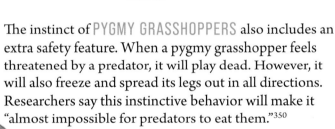

The instinct of PYGMY GRASSHOPPERS also includes an extra safety feature. When a pygmy grasshopper feels threatened by a predator, it will play dead. However, it will also freeze and spread its legs out in all directions. Researchers say this instinctive behavior will make it "almost impossible for predators to eat them."[350]

RESEARCHERS FOUND THAT 29 OF THE 50 DUCK SPECIES, INCLUDING THE FAMILIAR MALLARD, WILL PLAY DEAD WHEN THREATENED BY A PREDATOR. THEIR INSTINCT WILL HAVE THEM FLOP ON THE GROUND AND STOP MOVING. EVEN IF A FOX CARRIES THE "DEAD" DUCK TO THE FOX'S DEN, THE DUCK WILL CONTINUE BEING COMPLETELY LIMP, AS IF DEAD. THE RESEARCHERS CONCLUDED THAT THE DUCK IS CONTINUALLY MONITORING THE SITUATION — WAITING FOR THE RIGHT TIME TO RUN OR FLY AWAY. VETERAN FOXES LEARN THAT THEY NEED TO INFLICT FATAL WOUNDS EVEN IF THE DUCK APPEARS TO BE DEAD.[352]

DEAD OR ALIVE?

66

How would an octopus know to duplicate surrounding patterns to hide?

GROUND SQUIRRELS WERE GIVEN AN IMPRESSIVE ABILITY TO QUICKLY EVADE PREDATORS BY DARTING BACK AND FORTH WITH REMARKABLE SPEED AND AGILITY.

Squirrels Avoiding Predators: California ground squirrels' protective instincts include the action of playing dead. However, those instincts have several other actions that work well for this small animal that is prey for many larger animals.

When a squirrel detects a threat, it will quickly run up a tree. On the ground, it can run up to 20 mph, which is about the fastest that a human can run. Up a tree, it can move at up to 12 mph. Plus, once it is on a tree, the instinct will have it move to the back side of the tree and pull itself close to it. If necessary, the squirrel can even jump from tree to tree.

If there are no places to climb, the instinct will have the squirrel run very rapidly but in a zig-zag pattern with frequent sudden stops. You might wonder why the squirrel, running for its life, would keep suddenly stopping. Researchers have determined that this stopping is to look and listen for predators. Also, the instinct knows that most of the animals that want to eat squirrels have difficulty seeing them when they're not moving. All of these motions caused by the squirrel's instinct are effective in protecting it.[353]

However, the instinct uses its built-in knowledge about the animals that a squirrel should fear. The instinct knows that snakes are dangerous but slow. Mammals, like coyotes and bobcats, are faster but still not as fast as a squirrel. Flying predators require a rapid response by the squirrel to avoid being eaten.

When a snake is nearby, a squirrel may even go toward it. As the squirrel goes to a safe place, it will signal other squirrels of the danger with sound and by waving its tail.

When a coyote is nearby, a squirrel's instinct will generally have the squirrel run to the entrance of its burrow rather quickly. The squirrel will usually not slow down enough to signal danger visually, such as by tail wagging. It will probably only give the less useful danger signal of sound; sound warnings can't give other squirrels much information about where the danger is. Once the squirrel reaches its burrow, it will probably not go in immediately.

When an eagle is detected, a squirrel's instinct knows it cannot outrun it. This is when the zig-zag running pattern and sudden stopping will usually protect the squirrel. The instinct will also have the squirrel not just get to its burrow, but quickly get inside. The squirrel will usually give only a single, whistle-like warning for other squirrels.[354]

? Why do squirrels keep stopping when trying to escape?

There are other ways that God has given animals to escape from predators. Octopuses have the amazing ability to blend into their surroundings. If you've never seen this happen, you will probably have a hard time believing it's true. These creatures are able to change their appearance enough to prevent you, or a predator, from distinguishing between the creature and the surroundings.

Octopuses Blending into the Surroundings:

A marine biologist who has spent more than 20 years studying octopuses recently told of how, years ago, an octopus had moved, stopped, and disappeared. "I couldn't imagine how that animal could vanish in plain sight, not 3 feet away. **I'm still trying to figure out how they do it**" (emphasis added). He says that this amazing behavior is not a simple reflex; the creature is able to control it to some extent.[355] He thinks the octopus' instinct sees a surrounding pattern and uses about four basic colors and shapes to try to duplicate it.

God has designed such intricate means of protection for His most vulnerable creatures.

67

Why do chameleons have the ability to change the color of their skin?

Chameleons Blending into the Surroundings:

Chameleons, like octopuses, have the ability to change the colors and the patterns in their skin. Scientists debate why the instincts do this changing, but they are fairly certain how the instinct does it.

The surface layer of chameleon skin is transparent. Below that layer are several layers, each containing tiny sacs of pigment. As the instinct determines need, it can cause each tiny sac to empty its pigment for other creatures to see.

Each skin layer has only one pigment color. The bottom-most layer has brown pigment. On top of that is the layer with blue pigment. Next is the layer with yellow. Finally, just under the transparent layer is the layer with red. Blue, yellow, and red are the three primary colors that can be blended to produce all other colors. You can see that God has given the chameleon's instinct the ability to create almost any color by releasing the correct ones at the right locations so that they blend together in the sight of other creatures.[356]

Scientists know the details of how the instinct releases the pigments into the special cells (chromatophores) that hold the sacs. Chemicals will be sent from the nervous system and through the bloodstream to the chromatophores. As the pigment is released, these chromatophores enlarge. Thus, the chameleon can rapidly change its colors and their patterns.

Can you see it? Camouflaged in tree bark, an Eastern screech owl rests securely, utilizing its mottled plumage and earthy colors to evade larger predators while roosting. By blending seamlessly with its surroundings, it avoids detection from hawks and other birds of prey.

COLORATION IN CAMOUFLAGE?

Coloration can play a role in camouflage, as well as in something called thermoregulation. Darker shades of blue or aqua may help lizards absorb less heat from sunlight, assisting in temperature regulation in hot environments, while also breaking up their color patterns to make them harder to see by predators.

Lizard

CHAMELEONS ARE THE MASTER OF DISGUISE WHICH ARE ABLE TO BLEND INTO ANY BACKGROUND LIKE ROCKY WALL OR LUSH GREENERY.

The question of why a chameleon's instinct does this is debated by scientists. They all seem to agree that it is often done to either warm the chameleon (by using dark colors that absorb heat from the sun) or to cool off (by using light colors that reflect the heat away).

Some scientists think chameleons do not try to blend into their surroundings. These scientists will say chameleons already blend well into their natural surroundings. They have no need to blend in with their colors.

Other scientists think chameleons do try to blend into their surroundings. These scientists will give examples like how a chameleon will hide from a snake and a bird called a shrike. The instinct apparently knows the difference between these two animals that both want to eat chameleons. These scientists say that if a chameleon is threatened by a snake, it will do a simple act of becoming more pale or less intense in color. They say that is because the snake's vision is not great, and the snake will probably be looking up into the sun. Thus, the scene will be hard for the snake to see the chameleon above it. Becoming more pale is all that is needed.

FLYING SHRIKE

However, if the chameleon is being threatened by a flying shrike, the instinct will try to have the chameleon blend into the patterns of the ground below. The scientists say this is because the shrike has very good vision and will have the sun behind it. Blending into the ground patterns is necessary for survival.[357]

WHY DO SHRIKES IMPALE THEIR PREY? AFTER A SHRIKE KILLS ANOTHER ANIMAL, IT OFTEN IMPALES ITS PREY ON A THORN OR BARBED WIRE FENCE TO STORE IT FOR LATER. THE BIRD WILL THEN RETURN AND PICK AT ITS MEAL OVER THE COURSE OF SEVERAL DAYS. WAITING TO EAT PREY GIVES THE TOXINS FOUND IN SOME INSECTS TIME TO BREAK DOWN.

68

How do snakes use chemicals in their feces to smell like they're decomposing?

Animals have instincts that rapidly determine what to do when threatened. These instincts are so effective and complex that scientists call it a "defense system" that is "hard-wired"[358] such that the animal doesn't need to think about it. It is so well-designed that the animal will have the option of also using its conscious, thinking mind to determine how to react to a threat.

Generally, with animals being threatened, the animal's instinct will have to choose between freezing, fighting, or running.

FREEZING

FIGHTING

However, the instincts in many animals will also prepare the animal for fighting by creating analgesics (painkillers). These painkillers will facilitate the animal's escape or ability to fight by reducing the pain from any injury that might be inflicted by a predator or enemy. Scientists have studied how the instinct will use several parts of the brain to cause the body to generate and distribute these painkillers.[359]

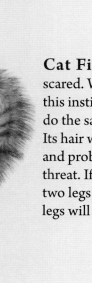

Cat Fight Instinct: You've probably seen a cat get scared. What you saw is what we all have seen, because this instinct is built into all domesticated cats. They all do the same thing. When scared, a cat will arch its back. Its hair will stand on end. Its ears will lower. It will hiss and probably swing one or two of its front paws at the threat. If it gets close enough, the cat will use its front two legs to hold it while the back two, more powerful legs will rapidly scratch the threat.[360]

Snake Striking and Biting: As is typical with most creatures, the fight strategy is usually the last resort for a snake. "Snakes will first hide, escape, or scare off predators before defending themselves through an attack."[361] Therefore, I want to at least mention some of the methods used by snake instincts to protect them.

If they can't hide or escape, snakes will coil their bodies. This could be in order to strike the threat, but usually it is to reduce their size and often to hide their head to protect it. While coiled, some snakes, for example rattlesnakes, will make noise with their tails to scare away the other animal or human. They will also usually hiss at the threat.

Some snakes will play dead. They may have their mouth open with their tongue hanging out. If picked up, they may roll back over onto their back.

A remarkable set of programming in some snakes takes advantage of a unique ability God put into their bodies. They have anal glands that can put out a foul-smelling chemical that smells like the snake is not only dead but already decomposing! The instinct's programming knows how and when to use this chemical. When there is no escape route, the snake's instinct will smear its body with that chemical and with its own feces. This trick is usually effective in protecting the snake.[362]

If they are forced to fight, snakes' instincts have a few tricks. They will usually coil their body to be able to jump farther. They will often flatten their head into a hood shape, like a cobra, to appear bigger and more threatening. They are very quick when they decide to strike and bite. However, their bite is not very effective on larger animals. This is partly because most snakes are not venomous to most creatures.[363] "Really, biting or striking a predator is a last resort for a snake."[364]

FLEEING

HIDE

KEELED SLUG SNAKE COILING UP AND HIDING IN LEAVES.

GET BIG

SPECIAL MUSCLES AND RIBS IN A COBRAS NECK SPREAD OUT TO FORM A "HOOD" WHEN HE FEELS THREATENED. THIS MAKES THE SNAKE LOOK BIGGER THAN IT REALLY IS AND MAY HELP SCARE PREDATORS AWAY.

ATTACK

RATTLESNAKE COILED FOR ATTACK, HISSING AND RATTLING

69
What would make badgers more afraid of humans than of bears?

"They are more afraid of you than you are of them" is often used to reassure hikers that even large predators, such as bears and lions, pose little threat to humans. This is common knowledge among biologists. They have guesses as to why.

However, THE BIBLE TELLS US THAT GOD SPECIFICALLY MADE IT THAT WAY, apparently to make life easier for humans to "HAVE DOMINION" OVER THE EARTH [GENESIS 1:28]. After the Flood of Noah's time, the humans and the animals got off the huge ark (very much like a super tanker of today), and God said the following: "Be fruitful and multiply, and fill the earth. And the fear of you and the dread of you shall be on every beast of the earth, on every bird of the air, on all that move on the earth, and on all the fish of the sea. They are given into your hand" (Genesis 9:1–2).

The idea that animals were created as vegetarians stems from the account of Creation in the Book of Genesis. In Genesis 1:29–30, God grants humans and animals the plants and fruits as food, indicating a vegetarian diet. Some interpret this to mean that all creatures, including animals, were initially intended to be herbivorous. The introduction of carnivorous or omnivorous behaviors came as a consequence of the Fall, as described in Genesis. After sin entered the world, the nature of animals, including their dietary habits, was altered. Isaiah (11:6–9) envisioned a future state of restoration and peace where predatory instincts and behavior cease, and animals return to a herbivorous lifestyle.

Here are some of the guesses of the biologists:[365]

Standing upright may scare the animals, since some 4-legged animals do this when attacking.
Humans usually go around in groups.
Humans have developed weapons that make us stronger than animals (this idea assumes the animals learn this, rather than have it in their DNA).
Today, the large animals are outnumbered by humans (this idea assumes the animals used to not be afraid of humans).
A new study "suggests that animals may be aware of the impact that humans have on their environments." (How could this be in their DNA?)

So, what was the "new study" of that last guess? The researchers used loudspeakers in a forest to play the sounds of various predators of badgers and of humans carrying on normal conversations. They used cameras to record the reactions of badgers to those sounds. Amazingly, the badgers were obviously more afraid of humans than of bears![366] Genesis 9 is very true!

Dog Baring Her Teeth and Growling: Like most animals, a dog would prefer to run away from any danger. But if she is trapped, her instinct will exhibit some of the following behaviors to protect her. Dog trainers have learned that dogs will generally follow a sequence of increasingly intense behaviors:

1	Becoming very still and rigid
2	Guttural bark that sounds threatening
3	Lunging forward or charging at the person with no contact
4	Mouthing, as though to move or control the person, without applying significant pressure
5	"Muzzle punch" (the dog literally punches the person with her nose)
6	Growl
7	Showing teeth
8	Snarl (a combination of growling and showing teeth)
9	Snap
10	Quick nip that leaves no mark
11	Quick bite that tears the skin
12	Bite with enough pressure to cause a bruise
13	Bite that causes puncture wounds
14	Repeated bites in rapid succession
15	Bite and shake

These instincts are obviously complex, and very effective at protecting a dog.

How to Calm A Dog's Instinct to Protect	
1	Respect the dog and its owner's space. Before you approach an unfamiliar dog, always ask the owner if it is okay to interact with their dog.
2	Keep your hands by your sides.
3	Listen to the dog's body language.
4	Don't indulge bad behavior.

70

How does the diving reflex work so seamlessly to help animals survive?

Cats Land on Paws

Many of God's creatures have what scientists call a "righting response," like cats, mice, birds, turtles, and certain amphibians, such as frogs and toads. When an animal is dropped in an upside-down position, it's instinct will try to turn it right-side up before it hits the ground. Cats are known for being remarkably adept at using the righting response. They always land on their paws.

The cat's sense of balance is similar to ours. Its inner ear detects changes in linear and angular acceleration. However, the cat's body is more flexible than a human body. So, when a cat is falling, the instinct "activates muscles throughout the body to flip the cat right-side up before it hits the ground."[369] This prevents the cat's body from being damaged as much as it might be by landing awkwardly.

GETTING STRAIGHTENED OUT: TURTLES HAVE A REMARKABLE, AND HANDY REFLEX, THAT HELPS THEM TO RIGHT THEMSELVES IF THEY ARE TIPPING OVER THIS BUILT-IN, GOD-GIVEN REFLEX IS ESSENTIAL, AS THE TURTLE WOULD HAVE NO RECOURSE IF IT ENDS UP ON ITS BACK.

Baby Ducks Follow Mother

When baby ducks come out of their eggs, they "know from the moment that they hatch that they are supposed to follow their mother. This is so that they stay safe, learn from her, and survive."[367]

This particular instinct is only partially programmed into the baby duck's genes. Studies have shown that the preprogramming will make the baby duck follow almost anything it sees immediately after hatching. One researcher imprinted the picture of a duck on the back of his boot and had newly hatched ducklings see that picture first. The result was that they followed his boot for the rest of their lives! The researcher concluded that the genetics would start the ducklings toward a close relationship with the mother and would then be "guided by **a mechanism** that under normal circumstances would not be corrupted by individuals of the wrong species"[368] (emphasis added). The researcher failed to see the wonder in God's creation!

DOLPHIN AND SEA LION UNDERWATER

OTTER

Animal Diving Reflex

This amazing reflex has been described earlier in this book, because God gave it to baby humans. This instinct makes important changes in the body as soon as the face is submerged in cold water. These changes give the baby a much better chance of survival under water, especially in deep water.

As a baby human grows up, this reflex becomes weaker. However, God has made it "particularly strong in seals, otters, and dolphins, which are mammals that spend a lot of time in the water."[370] As you would expect from a loving Creator, this "defense mechanism" is also in diving birds and reptiles that live in the water.[371]

It's another sign of a Designer when scientists find that this reflex is strong where it is most needed.

The diving reflex involves the coordination of various physiological systems, such as the respiratory, cardiovascular, and nervous systems. These systems work together seamlessly to facilitate the transition from land to water and ensure the animals' survival during dives. The integration and efficiency of these systems by the instincts are a powerful testament to God's design.

71

How do more than 5,000 species of birds journey thousands of miles every year?

From the precise hunting techniques of predators to the intricate nesting rituals of birds, adult animals rely on their instincts to navigate their environments, acquire food, reproduce, and ensure their survival. These instincts, deeply written into their genetic code by God, guide their actions and decision-making, enabling them to thrive in diverse habitats and successfully meet the challenges of their daily lives.

The study of adult animal instincts provides valuable insights into the remarkable strategies and adaptations that have allowed different species to flourish and maintain a delicate balance within ecosystems worldwide.

Without an atlas, road signs, or smartphone apps, more than 5,000 species of birds manage annual round-trip migrations. These journeys can be thousands of miles, with many birds often returning to the exact same nesting and wintering locations from year to year. But how do birds manage this amazing journey?"[372]

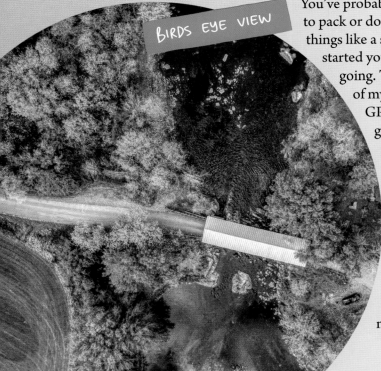

BIRDS EYE VIEW

You've probably been on a long trip and realized that you had forgotten to pack or do something before leaving. Over the years I've forgotten things like a special pillow, toothpaste, or tennis shoes. Before you started your trip, you needed to know when and where you were going. Thinking like an engineer, I did thorough planning for all of my trips. Most of them were taken before cells phones and GPS were available. Usually, all I had were maps and tour guide books. So, I would plan the routes and memorize the major turns I would need to make. I prided myself on my ability to remember maps and always know which way was north, thus never getting lost.

One particular trip took me onto Manhattan Island for the first time ever. My plan was to drive down from the north on the main highway, which becomes Broadway once you reach the main part of New York City. The northern part of Manhattan Island has few buildings and many trees. As I drove south, I enjoyed watching the number of buildings increasing, displacing the trees.

A FLOCK OF
SANDHILL CRANES

On this major thoroughfare, I began coming to intersections with other major roads. These intersections were the more modern roundabouts: instead of simple crossings with lights, they were small circles with entrances and exits for each of the four directions a car could go from the intersection. So, generally, I was to enter the small circle, pass the first exit, and take the second exit. This should keep me on the north-south Broadway, heading for the huge buildings of New York City.

So, I was enjoying watching the scenery changing. Suddenly, I realized that it seemed like the number of buildings was decreasing and the number of trees increasing! This made no sense. I then noticed that the sun was on the wrong side of my car! I was headed north, not south. Somehow, on one of the several roundabouts, I had gone all the way around the small circle and headed back the way I had come. I had gone so far in the wrong direction that I gave up completely the idea of going to downtown New York City.

So, my cerebrum failed to successfully guide me to the place I had seen on a map. Amazingly, God has put into the DNA and bodies of birds detailed instructions and abilities for taking very long trips to places they've never seen at all. These instructions include everything necessary for making a successful trip. They include preparation, knowing when to leave, knowing the route to take, knowing the best way to fly, knowing where to stop for food and water, and recognizing the final destination. He gave them physical abilities to do better than I did in staying on course while they are flying.

Biologists call these journeys "migrations." These migrations are very complicated and difficult for humans to understand. Let's look at some of the details of bird migrations.

ANNUAL MIGRATION OF ADÉLIE PENGUIN
IN SOUTH ORKNEY ISLANDS

Which Birds Migrate: "Roughly half of the world's nearly 10,000 known bird species migrate..." So, not all birds migrate. In fact, not all of the same species of bird will migrate. For example, some fraction of American robins will not fly south for the winter, though most do.[373]

72

What are the 12 diverse ways various birds are designed to migrate?

Migration is critical in the life cycle of birds, and without this annual journey many birds would not be able to raise their young. Birds **migrate to find** the richest, most abundant **food sources** that will provide adequate energy to nurture young birds. **If no birds migrated, competition** for adequate food during breeding seasons **would be fierce and many birds would starve.**" Instead, God gave them different migration patterns, times, and routes to give them and their offspring the greatest chance of survival. "Migration is a **dangerous but necessary journey** for many birds. Fortunately, they are **well equipped to survive the task…**"[374] (emphasis added).

The fact that large populations of birds generally live in their breeding ground for about six months before flying elsewhere for about six months, means that their food supplies in both places have about six months to replenish. This is a very smart system God has built into birds.

Why birds migrate is also related to the problem of predators. If all birds that live in an area stayed there all the time, the survival of bird chicks would be more tenuous. The animals that eat birds would multiply in that area, significantly reducing the bird population.

Migration means that soon after the chicks hatch, most of the bird population flies far away from that group of predators. "Many birds even migrate to specialized habitats that are nearly inaccessible to predators, such as steep coastal cliffs or rocky offshore islands."[375]

One other reason God has birds migrate is the problem of disease. Any large group of birds is susceptible to parasites and diseases that can decimate a bird population very quickly. Birds that fly away from the dead and dying through migration will have less chance of being affected.[376]

12. Drift
BLUETHROATS

9. Leap Frog
ROBIN

10. Reverse
BOHEMIAN WAXWINGS

7. Irruptive
RED BREASTED NUTHATCHES

6. Nomadic
BLUE JAY

5. Loop RUFOUS HUMMINGBIRDS

2. Latitudinal
RUBY-THROATED HUMMINGBIRD

MACAW DO NOT MIGRATE

11. Molt
BRITISH SHELDUCKS

3. Longitudinal
STARLINGS

8. Dispersal
EUROPEAN GREEN WOODPECKER

4. Altitudinal STORKS FROM NORTHERN EUROPE TO SOUTH AFRICA

VARIOUS WAYS BIRDS MIGRATE

12 Ways God Designed Birds to Migrate: When bird migration is mentioned, most people think of them going north in the spring, breeding, and then going south in the fall. This might be the most obvious kind of migration, but there are at least 12 different kinds that ornithologists (those who study birds) discuss. When you remember why birds migrate, it is logical that there could be several kinds of migration. Here is a list of 12 kinds:[377]

MOLTING BLUE JAY

1	Seasonal	This well-known migration is based between breeding and non-breeding places and occurs usually in spring and fall, but in some areas it is in wet and dry seasons.
2	Latitudinal	This is the most common type of migration and is north-south travel to avoid extreme cold.
3	Longitudinal	This is east-west travel common to birds in Europe and apparently caused by the effects of mountains on climate.
4	Altitudinal	This is where birds move up or down mountains, usually following the food supply.
5	Loop	This migration that God built into some birds takes advantage of different food sources in different locations at different times of the year. So, birds like the rufous hummingbirds will "follow a coastal route in spring on their way from Mexico to Alaska but take advantage of mountain wildflowers on an interior southbound route in autumn. Loop migration is also common with many seabirds and shorebirds as they use seasonal variations in wind patterns to aid their flight."
6	Nomadic	This migration is very erratic and depends on food and water resources. Generally, this migrating does not take the birds very far away from where they were.
7	Irruptive	This migration is similar to Nomadic Migration, except that the birds will travel much greater distances.
8	Dispersal	This migration is of birds that don't normally migrate. It is when young territorial birds are forced to leave where they were born, because the parents need that territory to survive. The young birds must go elsewhere and claim their own territory.
9	Leap Frog	As mentioned earlier, some species are divided between those that migrate and those that don't. If a migrating group comes upon a group of their own species who are not migrating, the migrating group will "leap frog" over them before landing. They will generally not mix together.
10	Reverse	This situation is an aberration where a few individuals, usually young birds, get confused and fly in the opposite direction of the rest of their species.
11	Molt	Some species of bird will fly to a safer location when they are molting (replacing old feathers to get new feathers, usually designed for the upcoming change of season) and less capable of flight.
12	Drift	This migration is rare, because it is when a species is flying a route not regularly used. It is usually caused by storms pushing the flock away from their regular migration route.

1. Seasonal
BAR-TAILED GODWIT

73

How could young common cuckoos find their way on a path they've never flown?

When Birds Migrate: Ornithologists are not certain how birds know when they should start migrating. They have several guesses. They think "changes in environmental conditions, such as the length of the day, may trigger migration by stimulating hormones, telling the birds it's time to fly." The birds may use "cues such as changes in light and possibly air temperature." God has created such remarkable programming in bird DNA that scientists call it a "bird's internal biological clock."[378] One writer said it can even detect "changing of the seasons based on light level from the angle of the sun in the sky and the overall amount of daily light."[379]

God's instinct system is so adaptable that "Local and regional weather conditions, such as rain, wind, and air temperatures can also influence decisions about when migratory birds take to the skies."[380] Another writer says the adaptability includes knowledge of food supply and if a bird needs time to recuperate from illness or injury.[381]

Even though the migration instincts are flexible, "most bird species follow **precise migration calendars….**" (emphasis added). This reflects the passage in Jeremiah 8:7 that refers to the birds knowing the time of their migration. More importantly, for the sharing of food resources, "those calendars vary widely for different species." Therefore, one writer says that "there are always birds at some stage of their journeys."[382]

Where Birds Migrate: Almost no matter where you live on this planet, if you are alert, you can witness birds migrating. This instinct causes birds of all sizes to go many different routes at many different times of the year. Usually these migrations are of large groups that sometimes include thousands of birds. Let's look at some of the routes.

God has built into each species a route that is designed to take them from a good breeding ground, where they've lived for about half a year, to a good place for growing strong the rest of the year. Then the instinct will return them to the breeding ground.

For each species, these routes have been designed for their flying ability. The many kinds of birds have differences that affect how well they can travel. The instincts take all of this into consideration for where and when each species will migrate. Important differences are things like flying speeds and how long they can fly without stopping. So, each species' route will include all necessary locations for resting and refueling.

One of the most important parts of my planning a long trip by car was making sure we had a motel room reserved and nearby restaurants. I did not want to be in an unfamiliar town with no place to sleep and eat. Well, amazingly, each bird species' route includes all the places necessary for resting and refueling, though many of the birds have never been there!

A RESEARCHER MEASURES A WILD WOODPECKER THE BIRD'S RIGHT LEG HAS A METAL IDENTIFICATION TAG.

ORNITHOLOGISTS

A CALIFORNIA CONDOR MARKED WITH WING TAGS, OFTEN USED TO STUDY MIGRATION.

Ornithology is a branch of zoology that concerns the "methodological study and consequent knowledge of birds with all that relates to them. The word "ornithology," comes from Latin *ornithologia* meaning "bird science."

Scientists have studied this and believe that young birds learn from their parents and from other adults some of the details of their species' route. "But what about inexperienced birds **migrating for the first time?** In one experiment, geographically displaced young common cuckoos **navigated back to roughly the same flight path** used by those birds that weren't displaced from their home"[383](emphasis added). So, as we've seen with many other instincts God has put into humans and animals, the basics of bird migration seem to be in the DNA, but they are adaptable to each situation.

THE ROUTES OF SATELLITE TAGGED BAR-TAILED GODWITS MIGRATING NORTH FROM NEW ZEALAND. THIS SPECIES HAS THE LONGEST KNOWN NON-STOP MIGRATION OF ANY SPECIES, UP TO 7,500 MILES (12,700 KM)

There are at least 12 different kinds of migrations, and there are always some birds in the process of migrating. Also, even if several species are flying a similar route, the different flying abilities mean different locations for resting and refueling. In other words, there is a huge variety of migration routes built into the various birds' DNA.

Here are some examples of amazing long-distance migrations. Arctic terns travel pole-to-pole roundtrips of over 60,000 miles. Even flightless birds migrate, such as the Adélie penguin, which makes a nearly 8,000-mile trek through frigid Antarctica. Scientists have recorded a bar-tailed godwit making a 7,500-mile, non-stop flight across the Pacific Ocean from Alaska to New Zealand over an 11-day period. Another non-stop flight is taken by great snipes from Europe to central Africa, covering 4,200 miles at speeds up to 60 miles per hour. Even the very tiny calliope hummingbird makes a 5,600-mile roundtrip migration.[384]

ARCTIC TERNS

CALLIOPE HUMMINGBIRD

These migratory routes show great variety, yet they are all obviously designed. Some will follow seacoasts. Others, if necessary and within the ability of the bird species, will cross dangerous deserts or long stretches of open sea. Some, again depending on the species' abilities, will include many stops while others will bypass all the stops. Yet most migratory routes will usually meet at logical places like mountain passes or narrow sea crossings.[385]

Some large barriers to birds are the Mediterranean Sea and the Sahara Desert. The bird instincts are amazing at deciding what each species is capable of doing. Some smaller birds will cross them non-stop and land exhausted in the first green space they find. More than 500 million birds cross the Sahara every year! Larger birds will usually go around these barriers, such as crossing at Gibraltar and going down the Atlantic coast.[386]

As stated earlier, no matter where you live, you have a good chance of seeing birds migrating. This is because the "whole world is criss-crossed with migration routes."[387] Birds in North America will generally fly south for the winter, even as far as South America. Birds in South America will fly north, even as far as Canada. In Asia, they will often go south to Indonesia or even Australia. In south Africa they will often go to central Africa or even Europe.

One researcher said, "I've been tracking birds for over two decades, and the ease with which birds seamlessly migrate between worlds is **absolutely astounding**"[388] (emphasis added).

74

What makes a hummingbird nearly double its body fat for migration?

How Birds Prepare for Migration: The actual flying of individual birds also requires the instincts to make a bird migration successful. Each bird needs to prepare, know how to navigate, know how to fly efficiently, and be prepared for problems.

How Birds Prepare for Migration: When the bird's biological clock determines that it will be migrating soon, the bird begins to prepare. Usually, the hormone levels will change, and the bird will build a greater fat supply for energy. A ruby-throated hummingbird can nearly double its body fat in just a week or two. "An osprey may take over two months to reach Africa. It can't store enough energy for non-stop flying, **because the extra fat would make it too heavy to fly**"[389] (emphasis added). Realize that an osprey could add the extra fat, but the instinct knows it wouldn't work as well as it does for the hummingbird. So, ospreys must stop many times to refuel, but the instinct knows that and includes the various stopping locations in the migratory route of ospreys.

"Once the birds are **in migration mode, a feeding frenzy ensues.** This allows the birds to accumulate fat to power their journeys," says Lucy Hawkes, a migration scientist at the U.K.'s University of Exeter who currently tracks Arctic terns. "**Somehow, [the birds] know** that they have to migrate soon and **get massive**"[390] (emphasis added).

The instinct will often cause molting: replacing of old feathers with new ones. "Many birds molt just prior to migration to take advantage of **more aerodynamic feathers that make flight easier and more efficient**"[391] (emphasis added). Notice how well God designed these instincts.

*"Even the stork in the heavens
Knows her appointed times;
And the turtledove, the swift,
and the swallow
Observe the time of their coming.
But My people do not know the
judgment of the LORD." (Jeremiah 8:7)*

How Birds Navigate: "One of the **greatest mysteries of migration is exactly how birds find their way** from one location to the next"[392] (emphasis added). Scientists have done many studies and found possible techniques but no clear explanation. Some of the techniques they think they've found are as follows:[393]

1	Magnetic Sensing	Biologists have found unusual chemicals, many containing very tiny iron particles, in the brains, ears, and even bills of many birds. They think they help the bird sense the earth's magnetic field. These probably act like an internal compass that the built-in programming of the instinct uses to orient the bird in the right direction on long journeys.
2	Geographic Mapping	Because each bird usually migrates on the same route several times during their lives, scientists think that the birds memorize major geographic features. So, while flying, they are guided by things like rivers, canyons, seacoasts, and mountain ranges.
3	Star Orientation	Biologists believe that the programming of the instincts can actually use the position of groups of stars (what we call constellations) to guide the bird to stay on the correct flight path.
4	Sun Orientation	The scientists also believe the instincts can use the position of the sun to help the bird follow the correct migratory route.
5	Learned Routes	Scientists know that in some bird species, such as swallows and Arctic terns, the young learn the correct route from their parents and older adults.

Ornithologists are learning that there may be other ways that birds navigate these long flight paths. The birds may use unique environmental sounds, like the sound of a waterfall, as a guide. They may consider the direction of a flock of birds of another similar species on their own migratory route.

Several scientists are becoming convinced that several bird organs help the bird have a "remarkable navigational ability." They've studied how the "bird's eyes interact with its brain" and concluded this probably helps the bird know which way is north. They've studied the "tiny amounts of iron in the neurons of a bird's inner ear" and believe that these act like a compass.

"**Most surprisingly,** a bird's beak" may help it navigate in at least two ways! These scientists think the bird's instinctive programming can create an "**olfactory map**" of the migratory route (emphasis added). The bird can use smells to guide it on its long journey.

They think the beak also may help the bird know how far north or south it is. They believe the trigeminal nerve going from the bird's beak to its brain is used to measure the strength of the earth's magnetic field. These scientists believe that the programming found in birds would not only know that this strength decreases as the bird flies away from the north pole (or south pole) toward the equator, but it would know exactly where on the bird's route would have that particular magnetic field strength![394] This is probably true, and it's amazing!

As one scientist concluded his report on bird navigation, he said, "however they do it, migration is a **truly incredible feat**"[395] (emphasis added).

TRIGEMINAL NERVE: A CRANIAL NERVE RESPONSIBLE FOR SENSORY FUNCTIONS, INCLUDING TOUCH, PAIN, AND TEMPERATURE SENSATION, AS WELL AS CONTROLLING THE MUSCLES INVOLVED IN BEAK MOVEMENTS.

75

Why do birds use their instincts to fly in V-formation?

How Birds Fly During Migration: When the birds actually start their migration, their instincts even change the way they fly. Many birds will fly higher than they normally do. This is because of several important things that God's programming knows. The air is cooler up high, and this will help keep their bodies from overheating due to the exertion. The wind patterns at higher altitudes will usually be in the desired migratory direction and will push the birds along.

When the wind direction happens to be against the bird, the instinct will make it fly low where obstacles like trees, mountains, and buildings will slow the wind speed.

Another change that the migratory instinct often makes is to change a daytime flyer into a nighttime flyer. Flying at night usually reduces the number of predators.[396] The night air is calmer, thus is easier to fly in, and is cooler, causing less evaporation of vital water.

The migratory instincts of many large birds have them fly by day in order to take advantage of the rising hot air ("thermals") that make it much easier for them to stay aloft. Many of the flying-insect eaters are also guided to fly during the day when those insects are flying.

For various reasons, the migratory instinct has birds fly together in large groups. One obvious reason is that there is safety in numbers. Predators are less likely to attack a group than one loner. Also, because they know each other and might even be related, when one bird cannot keep up due to sickness or injury, other birds will often stay with it until it is well.

While flying in large groups, the instincts use several ways of keeping them together. They can of course see each other, but the instinct will have many birds use sound, like geese honking, to have them stay together. With some birds, God has put on their rear ends markings easy to see, so that others in the flock can see them.

What Problems Can Occur During Migration: "Even with both physical and behavioral adaptations to make migration easier, this journey is filled with peril and there are many threats migrating birds face." Ornithologists estimate that more than 60% of some species never complete a full round-trip migration. You must remember that humans are continually changing what is on the migratory routes birds have used for thousands of years. Here is a list of some of the threats.[399]

1	Inadequate Food	subsequent starvation or lack of energy
2	Collisions	birds can fly into windows they can't see, or power lines, or wind turbine blades
3	Stopover Habitat Loss	due to human development or pollution
4	Predators	including wild animals, feral cats, and loose dogs
5	Poor Weather	storms that cause injury or disorientation
6	Light Pollution from Cities	the birds can't see the stars for navigation
7	Hunting	both legal and illegal

An amazing piece of aerodynamics that God built into bird migratory instincts has been adopted by the United States Air Force: flying in a V formation. You've probably seen a group of larger birds flying high in a V formation. Scientists have determined that "birds in V formation can fly 70 per cent farther than one bird flying alone." The front bird breaks up the wall of air, causing swirling air behind it. The swirling air is easier to fly through and can, at times, even flow in the direction of the bird's flight. The instinct even programs the birds to take turns leading the V formation, so that no one bird gets too exhausted.[397]

While flying, some birds are able to go much farther distances without stopping than other birds. Those birds are usually smaller and have gained much fat due to their migratory instinct. The birds that can't fly as far without stopping are usually larger. The instinct compensates for this by including resting and refueling stops at the right places along the planned migratory route.[398]

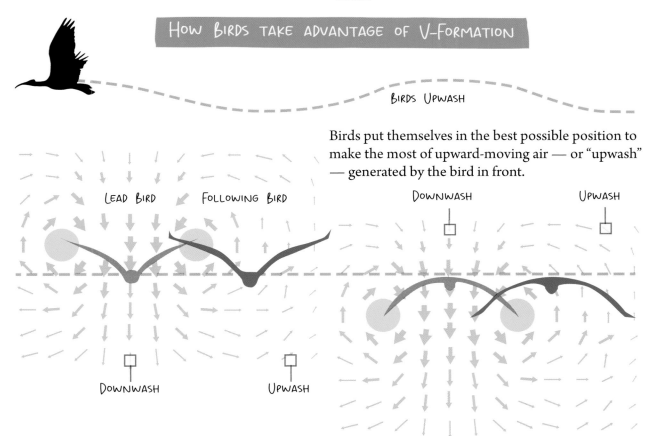

HOW BIRDS TAKE ADVANTAGE OF V-FORMATION

BIRDS UPWASH

Birds put themselves in the best possible position to make the most of upward-moving air — or "upwash" — generated by the bird in front.

LEAD BIRD   FOLLOWING BIRD

DOWNWASH   UPWASH

DOWNWASH   UPWASH

This quote sums up how well God has made bird migration:

"As scientists continue to unravel the **mysteries of bird migration,** the phenomenon remains **one of nature's great wonders. 'They're flying all night, feeding all day,** and doing it again,' Horton says. 'That's sort of **remarkable'**"[400] (emphasis added).

Flying in a V-formation creates an aerodynamic advantage for birds. As the leader bird cuts through the air, it creates an upwash of air that helps lift the following birds. This upwash reduces the effort needed for the trailing birds to maintain lift and, consequently, conserves their energy. What is called a downwash occurs at the end of the upwash. The birds in the V-formation, except for the lead bird, can take advantage of the reduced air resistance caused by the bird in front.

76

How does a caterpillar know to seal itself up in a cocoon?

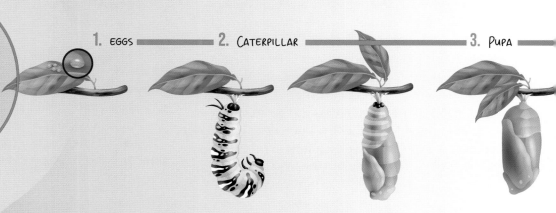

1. EGGS
2. CATERPILLAR
3. PUPA

The word METAMORPHOSIS has been used for 2,000 years. It basically means to change form or shape or to transform. Biologists use this word to describe how a caterpillar turns into a butterfly. Well, that word comes nowhere near describing the amazing changes that God's instincts in caterpillars are able to do! We'll look at the details in the next section.

Caterpillars are not the only animals that undergo metamorphosis. Biologists have found that "most insects undergo complete metamorphosis over the course of a lifetime."[401] This means that the insect will start in an egg. It will come out as a larva (most non-biologists would probably call it a worm). The larva spends almost its whole existence eating. This is because it is getting ready for the third "form" of its life: the pupa stage. The pupa does not eat at all. The pupa will surround itself with a hard shell which will protect it from predators. For some insects like moths, the hard shell will be a woven silk encasing that hardens. For others like butterflies, the larva will shed its skin for the last time, and the new exoskeleton will soon harden into what is called a chrysalis. Inside the cocoon or chrysalis, the pupa changes its form into the adult version of the insect (e.g., butterfly or moth). That adult version is the fourth and final "form" of the insect. [402]

Biologists have found that some fish also experience a similar change of form. The metamorphosis of salmon and of flounder will be discussed later.

Remember that these changes of form would not occur without programming (instincts). Listen to what biologists say about what they see. "Metamorphosis is a **remarkable** process…**astonishing**…. [S]ome scientists **believe** that the process of metamorphosis involves a sort of re-activating of **genes** that allow animal cells to change from one cell type to another" [403] (emphisis added). Notice several things in this quote. They are amazed by metamorphosis. They don't really understand it ("believe"). They are fairly certain that it is in the genes (built-in; preprogrammed); it is very hard to believe that there is any scientist that does not believe it is caused by the genes, they just don't know which ones.

CICADAS HAVE A LIFE CYCLE RANGING FROM 2 TO 17 YEARS, WHERE THEY WILL MOLT TO GROW INTO A FULL ADULT INSECT.

4. BUTTERFLY

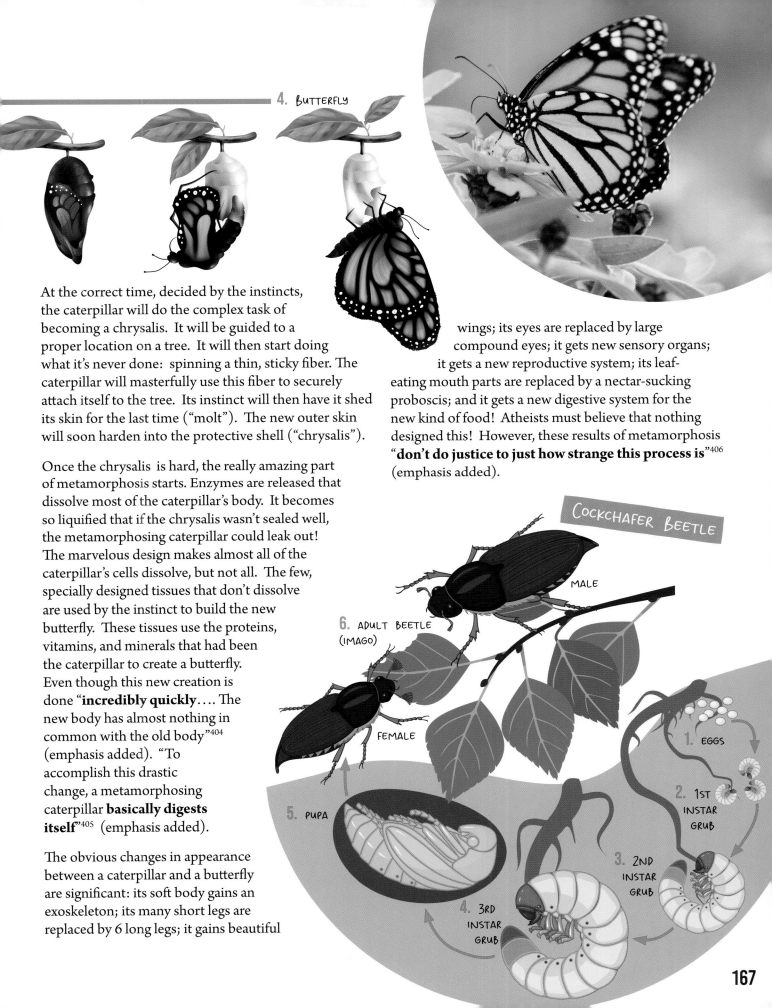

At the correct time, decided by the instincts, the caterpillar will do the complex task of becoming a chrysalis. It will be guided to a proper location on a tree. It will then start doing what it's never done: spinning a thin, sticky fiber. The caterpillar will masterfully use this fiber to securely attach itself to the tree. Its instinct will then have it shed its skin for the last time ("molt"). The new outer skin will soon harden into the protective shell ("chrysalis").

Once the chrysalis is hard, the really amazing part of metamorphosis starts. Enzymes are released that dissolve most of the caterpillar's body. It becomes so liquified that if the chrysalis wasn't sealed well, the metamorphosing caterpillar could leak out! The marvelous design makes almost all of the caterpillar's cells dissolve, but not all. The few, specially designed tissues that don't dissolve are used by the instinct to build the new butterfly. These tissues use the proteins, vitamins, and minerals that had been the caterpillar to create a butterfly. Even though this new creation is done "**incredibly quickly**…. The new body has almost nothing in common with the old body"[404] (emphasis added). "To accomplish this drastic change, a metamorphosing caterpillar **basically digests itself**"[405] (emphasis added).

The obvious changes in appearance between a caterpillar and a butterfly are significant: its soft body gains an exoskeleton; its many short legs are replaced by 6 long legs; it gains beautiful wings; its eyes are replaced by large compound eyes; it gets new sensory organs; it gets a new reproductive system; its leaf-eating mouth parts are replaced by a nectar-sucking proboscis; and it gets a new digestive system for the new kind of food! Atheists must believe that nothing designed this! However, these results of metamorphosis "**don't do justice to just how strange this process is**"[406] (emphasis added).

COCKCHAFER BEETLE

MALE

6. ADULT BEETLE (IMAGO)

FEMALE

1. EGGS

2. 1ST INSTAR GRUB

3. 2ND INSTAR GRUB

4. 3RD INSTAR GRUB

5. PUPA

77

Why does the instinct of a frog direct certain cells to shred their DNA and die?

1. EGGS   2. TADPOLE

Tadpole to Frog: The metamorphosis of a tadpole into a frog is not as complete as the metamorphosis of a butterfly. Instead of the whole body of the caterpillar being dissolved and converted into a butterfly body, only certain parts of a tadpole are converted into frog parts.

The instinct orders tadpole cells that won't be needed in the frog to "**shred their DNA and die.** The dead cells are **then cannibalized** for energy and raw materials…"[407] (emphasis added). The instinct converts cells in the tadpole tail into legs of the new frog. It converts the tadpole gills, necessary for breathing underwater, and other parts into air-breathing lungs for the frog.

This instinct is able to TAKE A TADPOLE THAT LIVES UNDERWATER eating algae and plants and TURN IT INTO A FROG THAT LIVES ON LAND breathing air and eating insects.

It's worth noting the way the evolutionists worded their description of the metamorphosis of the frog: "tadpoles '**order**' the cells they don't need anymore to shred their DNA and die."[408] They strongly imply that the tadpole "consciously" does this amazing action. That is a lie! It is also typical of the way evolutionists talk about many of the instincts described in this book.

EUROPEAN COMMON BROWN FROGS — WITH EGGS

3. TADPOLE WITH 2 LEGS   3. TADPOLE WITH 4 LEGS   4. FROG WITH SHORT TAIL   5. ADULT FROG

EGGS

NEWLY HATCHED

JUVENILE EUROPEAN FLOUNDER

ADULT FLOUNDER

Flounder: Newly hatched flounder look like most other fish. They are thin and swim vertically compared to the sea bottom. This allows them to swim fast and see everything around them. However, early in their lives the metamorphosis instinct takes over.

The young fish essentially begin living on their side. "Through cellular changes, the eye and nostril from [one] side actually migrate to join the other eye and nostril on what is now the 'top' side of the fish."[409] Adult flounders are rather strange looking. God has designed them to lay on the bottom of the ocean, mainly waiting for prey to swim by. So, they lie flat on the sea bottom and have both eyes on the "top" side.

Notice how the following quote about metamorphosis sounds like a literal fulfillment of Romans 1:20–21 which says, "For since the creation of the world His invisible attributes are clearly seen… even His eternal power and Godhead… they did not glorify Him as God…." The authors of much of what is quoted in this book about the salmon, flounder, frog, and butterfly life cycles said this, "**Evolution** sure has some **creative ways** of doing things!" (emphasis added). God certainly does!

169

78

How do freshwater salmon develop new organs to cope with salt water?

Salmon: Salmon eggs are laid in freshwater rivers. The baby salmon (called a "fry") hatches into freshwater. For reasons unknown to biologists (but they know there must be a reason), the salmon needs to go out into the ocean to live most of its life. This is a major problem. Why? Freshwater fish will die if put into the ocean!

"All organisms must regulate their salt/water balance. This is why humans can't drink seawater without dying: the salt would overwhelm our cellular chemistry, and our cells would not function properly. In just the same way, freshwater fish typically cannot live in saltwater. To become saltwater fish, then, salmon **must develop new organs and cellular mechanisms to cope with the salt water**"[410] (emphasis added).

This amazing change in physiology is just part of the work of instincts in controlling the life of a salmon.

When a fry is hatched, instincts immediately give it the desire and guidance for survival, including finding and eating tiny invertebrates and even the dead bodies of adult salmon. "Fry **instinctively** hide, deal with river currents, learn to school together and many other survival skills"[411] (emphasis added).

Fry will live in the fresh water anywhere from a few days to two years, depending on the species. Then the instincts will start the metamorphosis of their bodies so that they can live in the saltwater ocean. At the proper time, the instinct will make the salmon migrate down river to the ocean.

Knowing that everything in nature appears to have a purpose, biologists guess at the reason for salmon starting out in fresh water but moving to salt water. The guess repeated by many biologists is that it increased the amount of available food. The young salmon live off of what is in the freshwater river, while the older salmon live off of what is in the saltwater ocean. This seems more likely to fit a pattern created by a Designer.

Once in the ocean, the salmon will typically travel many hundreds of miles to the feeding ground used by their species. How do these salmon know where previous generations lived in the vast ocean? Remember that they never met their parents or probably any of the salmon that had been to those feeding grounds!

During their time in the ocean, salmon will move around for one to five years and to specific locations, depending on their species. "Scientists believe that salmon **navigate by using the earth's magnetic field like a compass**"[412] (emphasis added). All of this has to be in their genes, preprogrammed. They spend this time eating and storing energy for the difficult final stage of their lives, the egg production ("spawning").

When it's time to lay their eggs, salmon will begin swimming toward the same river they were born in. They will stop eating before they get to the fresh water. They will expend a great deal of energy swimming upstream, usually going many miles up the river. When they reach the place they were hatched, they will lay and fertilize their eggs.

After this, the emaciated, worn-out adults will soon die. Biologists don't appear to be sure why most salmon die at this point. There are some salmon that return to the ocean and spawn again later. The guesses for why so many die after the first spawning is the combination of not eating while swimming up the river, the enormous amount of energy used in the swimming, and the effect of being in fresh water with a body designed for salt water.

AN ALASKAN BROWN BEAR TRIES TO CATCH SOCKEYE SALMON IN ALASKA

It's interesting that biologists are so used to seeing a purpose for EVERYTHING in nature (which is what you would expect of a Creator, but NOT what you would expect if it happened by dumb luck) that some see the lack of eating in the last few weeks as a benefit for egg production: "It's digestive system will atrophy and shrink making more room for eggs and sperm."[413]

EGGS	FEMALE SALMON LAY EGGS IN GRAVEL NESTS (REDDS) AND MALES FERTILIZE THEM
ALEVINS	YOUNG SALMON THAT HATCH FROM EGGS WITH A YOLK SAC FOR NOURISHMENT.
FRY	SALMON FRY ARE FREE SWIMMERS AFTER ABSORBING THE YOLK SAC.
SMOLT	SMALL SALMON UNDERGO SMOLTIFICATION TO ADAPT TO SALTWATER ENVIRONMENTS.
ADULT	SMOLTS GROW INTO FULL ADULT FISH AND SPEND YEARS IN THE OCEAN.
SPAWNER	ADULT SALMON SWIM UPSTREAM TO REPRODUCE AFTER DEVELOPING REPRODUCTIVE ORGANS.

FRESH WATER RIVER

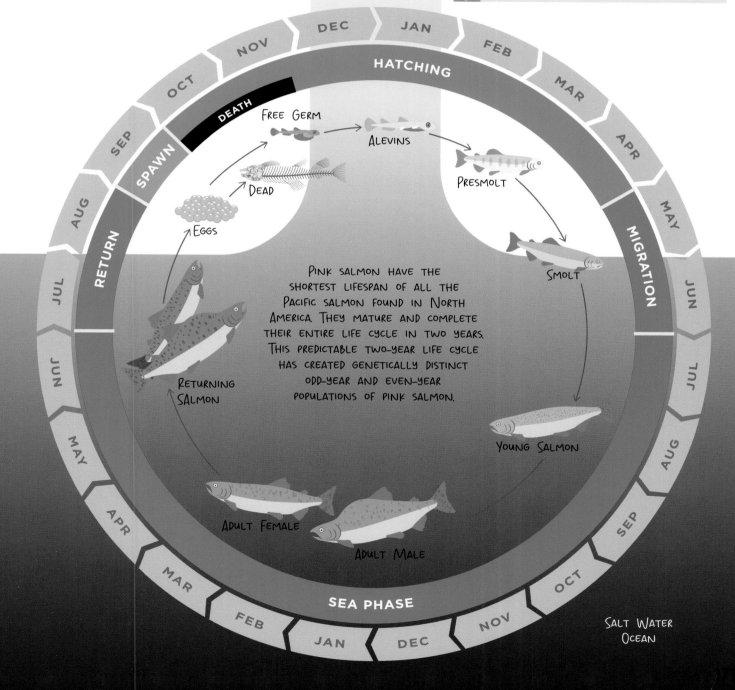

PINK SALMON HAVE THE SHORTEST LIFESPAN OF ALL THE PACIFIC SALMON FOUND IN NORTH AMERICA. THEY MATURE AND COMPLETE THEIR ENTIRE LIFE CYCLE IN TWO YEARS. THIS PREDICTABLE TWO-YEAR LIFE CYCLE HAS CREATED GENETICALLY DISTINCT ODD-YEAR AND EVEN-YEAR POPULATIONS OF PINK SALMON.

SALT WATER OCEAN

79

What makes a sea slug cut off its own head when infected with a parasite?

Stretching: Have you seen a dog or a cat wake up from a nap and yawn and stretch? You may have heard that you also need to stretch after sleeping. You especially need to stretch your muscles before doing any significant exercising. Well, God put that knowledge into most animals. Their instincts make them stretch when they wake up.

Cats "and dogs always stretch out their spine and extremities" when they wake up. This prepares their muscles and joints for their normal activities. Stretching "maintains flexibility, keeps fluids circulating and prevents injury."[416]

ELYSIA ORNATA
SEA SLUG

Slugs Decapitate Themselves When Infected: God has created several ways for humans and animals to deal with infections by parasites. Human instincts will usually increase the body's temperature (get a fever) and special blood cells (killer T cells, macrophages, etc.) attack the invaders. Some animal instincts will make the animal eat certain plants that contain chemicals that will kill the invaders. One of the strangest instincts is that of at least two species of sea slug (they look like worms or snails without their shells).

"Researchers were **astonished** to observe slugs in captivity **cutting off their own heads** after their bodies became infected with parasites."[414] Can you imagine someone suggesting that you cut your head off to get rid of parasites in your body? That's insane! Yet, that is what the instincts of those sea slugs make them do. The instinct knows that those sea slugs (unlike almost every other creature on this planet) can regrow their whole body from just their head. It takes only about 3 weeks. The new body will be parasite-free.

Note that "the bodies never grow back new heads."[415] How do their instincts know the heads can grow new bodies? How do the instincts actually make the new bodies grow out of the heads? This programming must be amazing. This programming by the Creator is just amazing!

DUNG BEETLES NOT ONLY LIVE IN DUNG, BUT THEY ALSO BREED THERE AND EAT IT. WE'RE SPECIFICALLY DISCUSSING DUNG OR POOP HERE. THESE CREATURES PERFORM ALL THESE ACTIVITIES WITHIN FECES, AND THEY EVEN TAKE IT A STEP FURTHER BY ROLLING BALLS OF DUNG WHILE STANDING ON THEIR HEAD AND MOVING BACKWARDS USING THEIR HIND LEGS.

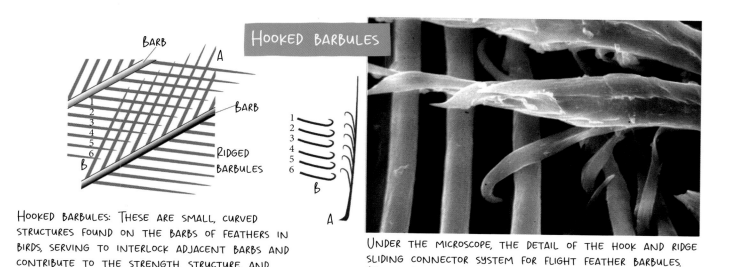

HOOKED BARBULES

HOOKED BARBULES: THESE ARE SMALL, CURVED STRUCTURES FOUND ON THE BARBS OF FEATHERS IN BIRDS, SERVING TO INTERLOCK ADJACENT BARBS AND CONTRIBUTE TO THE STRENGTH, STRUCTURE, AND FLEXIBILITY OF THE FEATHER

UNDER THE MICROSCOPE, THE DETAIL OF THE HOOK AND RIDGE SLIDING CONNECTOR SYSTEM FOR FLIGHT FEATHER BARBULES. (Photograph courtesy of Prof. David Menton, Washington University School of Medicine, St. Louis, Missouri)

How Birds Stay Warm: God has given most birds two primary kinds of feathers. The main kind are used to enable the bird to fly. This kind of feather is "a **magnificent adaptation** for flight" and is "**remarkably** light and strong." Most people don't know how marvelously it works on a bird wing. Due to an "**intricate system** of tendons … when the wing is raised they open like the vanes of a blind, greatly reducing resistance, but close completely on the downstroke, thus **greatly improving the efficiency** of flight"[417] (emphasis added).

The second main kind of feather is called down. It is designed mainly for keeping the bird warm. When the bird is experiencing cold weather, its instinct guides it to fluff up those down feathers. Over a very short time, the trapped air inside those fluffy feathers will be heated up by the bird's body. "Birds' feathers provide **remarkable insulation** against the cold…"[418] (emphasis added).

DOWN FEATHERS TO KEEP WARM

As usual, the instinct is more complicated than what has been described. God added at least two more behaviors to protect the bird. When it is bitterly cold, the instinct will have the bird lift one leg and pull it inside the warm down feathers. It will also have the bird turn its head almost 180 degrees and bury its beak into the warm feathers.

80

Why do certain warm and cold-blooded creatures go into a kind of dormancy?

Many animal life cycles include a time of being dormant. It is a time when the animal not only stops growing but almost appears dead. Usually, the breathing rate and heart rate will become very slow. The animal will stop eating. These periods of dormancy can last for months. Scientists have concluded that animals go dormant to conserve energy and survive bad weather.[419]

Researchers have determined that there are three basic kinds of animal dormancy. Researchers actually have at least four names and admit there is overlap. Also, there are many examples of animals going dormant when circumstances indicate they will be safer being dormant. I will present only the three easily defined and common kinds: hibernation, brumation, and estivation. No matter the kind of dormancy, "animals have an **amazing ability** to adapt and survive in their different environments"[420] (emphasis added).

Hibernation: Hibernation is what some warm-blooded creatures do every winter. Before the cold weather starts, the instincts of those animals are causing them to get ready for it. For example, in late summer, brown bears will all go down to the rivers for "one last big feed before hibernation."[421] The instincts know that food will be scarce and extreme weather will make survival difficult. God has put the necessary programming into these animals in order to accomplish everything necessary for survival.

Hibernation is a **complex set of behaviors.** It requires being aware of the time of year, engaging in pre-hibernation preparation tasks, hibernating, and emerging from hibernation **at the right time; quite a complex process,** to say the least[422] (emphasis added).

Hibernation requires many physiological changes in the animal, such as decreasing the heart rate (by as much as 95%) and lowering the body temperature. This can be dangerous in freezing weather, so God has some of them use a unique method of generating heat: "a **regulated process** in which the proton gradient generated by electron transport in mitochondria is used to produce heat instead of ATP…"[423] (emphasis added). Basically, animals use a process where the energy produced by their cells is used to create heat instead of the usual energy called ATP.

Hibernation is not really sleeping. It is more extreme than sleeping. Even the metabolism slows down to less than 5% of normal. Some physiological functions slow considerably, such as heart rate and breathing.. When sleeping, the unconscious mind is still working. One researcher "found that hibernators have to undergo periodic arousals so they can catch some Zs!"[424]

BLACK BEAR HIBERNATING

Aestivation: Aestivation (or estivation) is when an animal goes dormant when the environment is too hot or dry. Vertebrates and invertebrates have this instinct. This dormancy is similar to hibernation. A major difference is that it appears to completely depend on the situation rather than time or season. For example, a snail that is aestivating can come completely out of it in just 10 minutes. Generally, though, the physiological changes are like those of hibernation.[427]

FOREST TURTLE AESTIVATING UNDER PINE NEEDLE NEST ON A HOT SUMMER DAY.

WOOD FROGS GO ONE STEP FURTHER THAN THE MAJORITY OF ANIMALS BY FULLY CEASING TO BREATHE AND HAVING A BEATING HEART.

GARTER SNAKES COMING OUT OF HIBERNATION AND INTO TENS OF THOUSANDS STRONG MATING BALL IN SOUTHERN CANADA

Brumation: Brumation is basically the hibernation for cold-blooded animals. What happens to the reptiles that brumate is very similar to what happens to the mammals that hibernate. A few differences are that the reptiles will still need to drink water and will occasionally move around.[425]

This "amazing feat" is "built into their instincts and DNA." Some salamanders can go 10 years without eating! Some other reptiles turn their heart rate down to only 1 beat per minute! The scientist who said that, though he would not give God the glory, had a very informative summary of this instinct God has given many animals:

A species' **physiology is totally adapted** to its typical food intake. If its normal feeding cycle has long gaps (hibernation, mating period, migration, etc.), then the **physiological adaptations include** big fat stores, lowered metabolism, breakdown and use of muscle, stuff like that. That's all about calorie intake and energy use. Any species that has long gaps in feeding **must have** these kinds of **physiological tricks**[426] (emphasis added).

SLEEPING ARGENTINE BLACK AND WHITE TEGU (LIZARD)

175

81

How does a chemical exuded by dead bees drive the hive instinct to remove them?

Beehive Hygiene: We probably never think about it, but animal homes need to be cleaned just like ours. This is especially true if the "trash" is dead bodies. They will decay and usually become toxic for those still living. God of course thought of this need and created an instinct to make animals clean their homes.

One example of this housekeeping instinct is seen in bees. A beehive is full of thousands of bees. So it often happens that a bee will die inside the hive. The living bees will dutifully pick up the dead body and carry it outside the hive. The instinct even guides the living bees to correctly deal with a larva that has died inside its cell: the bee will cut open the cell and then carry the dead larva outside the hive.

BEES WORKING HIVE

"While this **appears to be an intelligent response** to the possibility of contamination in the hive, it is **in fact a blindly programmed reaction** to the presence of a chemical, oleic acid, exuded by dead bees"[428] (emphasis added). Researchers have found that this instinct is controlled by two genes. One guides the bee to cut open the hive cell, while the other gene guides the bee to pick up and carry the dead body outside the hive.

However, we know that God does nothing blindly. He has designed this activity for the health of the hive.

OTHER INSECTS, LIKE ANTS, EXHIBIT CLEANING BEHAVIORS WITHIN THEIR COLONIES. THEY REMOVE DEAD MEMBERS, DEBRIS, AND WASTE TO MAINTAIN A CLEAN AND HEALTHY LIVING ENVIRONMENT.

Herd Instinct: The herd instinct goes directly against the theory of evolution. Evolution theory is based on the idea of survival of the fittest. Yet, the instinct God put into some animals makes them behave unselfishly.

One scientist watched an elephant herd in Kenya and noticed a teenaged elephant that was unable to keep up with the herd. He learned that it had been a cripple for several years. He saw that the herd would occasionally stop to allow the crippled elephant to catch up. He even saw the leader of the herd go back and feed the crippled one. The scientist pointed out that the herd members had nothing to gain by helping the young elephant. Nevertheless, they adjusted their behavior to allow the cripple to remain with the group.[429]

This instinct is in several different kinds of creatures, including wildebeest, zebras, and bison.

TWO BABY ELEPHANTS INTERACT WHILE AN ADULT GENTLY TOUCHES THEM WITH HER TRUNK ELEPHANTS ARE KNOWN FOR THEIR STRONG FAMILY BONDS AND CARING FOR EVERY MEMBER OF THE HERD.

God's Design in Instinctive Behavior:

Instincts are very much like computer programs designed to do important actions for God's creatures. This book presented many examples of those actions in humans. Before doing that, we should look at how these programs work.

To summarize, these programs are in the DNA of each baby human. As the baby develops and is finally born into the world, these programs put instructions into his nervous system. These instructions use the nervous system to control various body parts to accomplish the "instinctive behavior."

What is the nervous system? "The nervous system is an **extraordinarily complex** communication system that can send and receive voluminous amounts of information simultaneously"[430] (emphasis added). It is the brain (cerebrum, cerebellum, and brainstem), spinal cord, and nerves that branch out to every part of the human body. Each of these nervous system parts are made of tiny cells called neurons. Each neuron has many dendrites (branches that resemble tree branches) and one long arm (called an axon) that ends in more dendrites. Amazingly, the axons vary in length from less than one millimeter to about one meter.[431]

The nervous system is essentially an electrical system. The neurons connect to each other and to body parts with their many dendrites, like electrical wires in a house. When necessary, electrical signals go into a neuron from a dendrite connection (a synapse) and out through the axon and its dendrites to other neurons or body parts. A single axon with all of its dendrites can innervate multiple parts of the brain.[432]

The nervous system has an almost unimaginable number of parts. Just the brain contains about 100 billion neurons. The total length of those neurons in the brain is more than the distance from the earth to the moon! The only way the nervous system can work is if there are continuous, CORRECT, connections between neurons and body parts. Imagine what would happen to your computer if just a few of the electrical connections inside were not made correctly or were not connected. It probably would not work at all. Well, your computer is simple compared to the human brain. "Nothing in biology is more ordered than the hardwiring of our brain." Its hardwiring makes it able to "achieve the top performance required to equip us with this 'cognition computer.'"[433]

The "wiring" and connections are more complicated than already stated. For example, the "connections" (synapses) are not physical connections. They are places where the electrical signal causes ions (sodium, potassium, and calcium) to travel from one neuron's dendrite to another neuron's dendrite and cause the electrical signal to start in the new neuron. This is how the electrical signal moves from any of the body's many receptors and through the nervous system.[434]

So, how does an instinct use the nervous system? Typically, when an instinct operates, this is what happens. A body part, or parts, will send electrical signals to neurons. They will pass them on until they find the program of instructions (the instinct). This could be in one of many places in the nervous system: a large nerve, the spinal cord, the cerebrum, the cerebellum, but usually in the brainstem. These instructions will send electrical signals to neurons which will activate the correct body parts.

Conclusion: My prayer is that this book has made you better appreciate the amazing body God has given to you and to the animals around you.

Realize that this book focused only on the programming God created for us and the animals around us. It did not deal with the many obviously designed physical features of humans and animals. For example, anatomists have written many books about the human hand. That one small part of the human body is so remarkable that one scientist wrote a whole book dealing with only the hand (properly titled *The Hand*).

As quoted at the beginning of this book, *Psalm 139 verse 14 says, "I will praise You, for I am fearfully and wonderfully made; marvelous are Your works, and that my soul knows very well."*

The key question for someone doubting there is a God is whether these complex sets of instructions could have appeared without a Programmer.

ENDNOTES:

1. "Brain Anatomy and How the Brain Works" in Health, at Johns Hopkins Medicine, https://www.hopkinsmedicine.org/health/conditions-and-diseases/anatomy-of-the-brain.

2. The Free Dictionary by Farlex, Inc. Version 11.1 on TheFreeDictionary.com.

3. Bryan Johnson, "Top 10 Human Reflexes and Natural Instincts," January 28, 2012, https://listverse.com/2012/01/28/top-10-human-reflexes-and-natural-instincts/.

4. Letter from Albert Einstein to Erik Gutkind in 1952, quoted in "Einstein, God, and Religion: What He Thought and What He Believed," Union of Catholic Christian Rationalists, https://www.uccronline.it/eng/2018/08/14/einstein-god-and-religion-what-he-thought-and-what-he-believed/.

5. Mark S. Blumberg, "Development Evolving: The Origins and Meanings of Instinct," https://www.ncbi.nlm.nih.gov/pmc/articles/PMC5182125/#R17.

6. Timothy D. Johnston, Laura Edwards, "Genes, Interactions, and the Development of Behavior" in Psychological Review, 2002, Vol. 109, No. 1, 26–34, by the American Psychological Association, Inc., https://core.ac.uk/reader/192385186?utm_source=linkout.

7. "Innate & Learned Behavior," chapter 29, Lesson 1, https://study.com/academy/lesson/innate-learned-behavior.html.

8. Dr. Brenda Shook, National University, "Ask an Expert: Is Human Behavior Genetic or Learned?" https://www.nu.edu/resources/ask-an-expert-is-human-behavior-genetic-or-learned/.

9. Praveen Shrestha, "Instinct Theory of Motivation," Motivation Emotion, November 17, 2017, https://www.psychestudy.com/general/motivation-emotion/instinct-theory-motivation.

10. Lebogar, "What Is the Meaning of Animal Instinct?" https://www.enotes.com/homework-help/what-meaning-animal-instinct-150131.

11. "Innate Behavior of Animals," https://bio.libretexts.org/Bookshelves/Introductory_and_General_Biology/Book%3A_Introductory_Biology_(CK-12)/10%3A_Animals/10.04%3A_Innate_Behavior_of_Animals, accessed August 14, 2020.

12. Marion Sipe, "What Is Innate and Learned Animal Behavior?" https://sciencing.com/innate-learned-animal-behavior-6668264.html.

13. James Beckerman, MD, FACC, "What to Do If Your Heart Races, Slows Down, or Skips a Beat," August 24, 2020, https://www.webmd.com/heart-disease/atrial-fibrillation/what-to-do-heart-races#1-6.

14. Ibid.

15. "The Brain Stem," Boundless Anatomy and, https://courses.lumenlearning.com/boundless-ap/chapter/the-brain-stem/) on https://courses.lumenlearning.com/boundless-ap/chapter/the-brain-stem.

16. "Control of Blood Pressure," https://courses.lumenlearning.com/boundless-ap/chapter/control-of-blood-pressure/.

17. Mather Hospital Northwell Health, "Cardiology 101: What Does Blood Pressure Really Mean?" https://www.matherhospital.org/wellness-at-mather/diseases-conditions/cardiology-101-what-does-blood-pressure-really-mean/.

18. "Control of Blood Pressure," https://courses.lumenlearning.com/boundless-ap/chapter/control-of-blood-pressure/.

19. Ibid.

20. Kathryn Watson, "How Do Babies Breathe in the Womb?" March 13, 2017, https://www.healthline.com/health/pregnancy/how-babies-breathe-in-the-womb.

21. Ibid.

22. "Respiratory System" WebMD, https://www.webmd.com/lung/how-we-breathe.

23. Bradley J. Fikes, "Body Parts: Pleural Cavity Maintains a Vital Vacuum," June 17, 2007, San Diego Union-Tribune, https://www.sandiegouniontribune.com/sdut-body-parts-pleural-cavity-maintains-a-vital-vacuum-2007jun17-story.html.

24. Ibid.

25. "What Happens When You Cough?" https://www.robitussin.com/cough-cold-center/anatomy-cough/.

26. "Stroke Education," https://www.strokeeducation.net/.

27. Kimberly Holland, "Why Do We Sneeze?" September 18, 2018, https://www.healthline.com/health/why-do-we-sneeze.

28. Mason Weupe, Jacelyn E. Peabody Lever, "Moving Mucus Matters for Lung Health," September 11, 2019, https://kids.frontiersin.org/article/10.3389/frym.2019.00106.

29. Ibid.

30. Ibid.

31. Ibid.

32. David Reuben, Everything You Wanted to Know About Nutrition (New York: Simon and Schuster, 1978), p. 79.

33. David E. Leib, "Thirst," https://www.ncbi.nlm.nih.gov/pmc/articles/PMC5957508/.

34. Ibid.

35. Sharon Palmer, RD, "Taking Control of Hunger — Lessons on Calming Appetite and Managing Weight," Today's Dietitian, April 2009, https://www.todaysdietitian.com/newarchives/040609p28.shtml.

36. Tracey Roizman, D.C., "How Does Your Stomach Tell Your Brain That You're Full?" https://www.livestrong.com/article/489875-how-does-your-stomach-tell-your-brain-that-youre-full/.

37. DMC Detroit Medical Center, https://www.dmc.org/services/pediatrics/pediatric-healthy-living/corporate-content/newborn-reflexes.

38. Lynne Eldridge, MD, "Function and Disorders of the Alveoli, Minute Structures of the Lung Vital to Respiration," medically reviewed by Doru Paul, MD, on July 20, 2020, https://www.verywellhealth.com/what-are-alveoli-2249043.

39. Kathryn Watson, "How Do Babies Breathe in the Womb?" March 13, 2017, https://www.healthline.com/health/pregnancy/how-babies-breathe-in-the-womb.

40. From "1.4 The Somatic Nervous System" in Neuroscience: Canadian 1st Edition Open Textbook, by Pressbooks, http://neuroscience.opentext.utoronto.ca/chapter/anatomy-physiology-the-somatic-nervous-system/.

41. "Nutritional Requirements throughout the Life Cycle," https://nutritionguide.pcrm.org/nutritionguide/view/Nutrition_Guide_for_Clinicians/1342043/all/Nutritional_Requirements_throughout_the_Life_Cycle.

42. "Benefits Of Breastfeeding — Incredible Facts About Breastfeeding for Mommy & Baby," http://www.breastfeedingrose.org/incredible-facts-about-breastfeeding-for-mommy-baby/.

43. David Reuben, M.D., Everything You Always Wanted to Know About Nutrition (New York: The Maleri Corp., 1978), p. 85–86.

44. Paul A.S. Breslin, "An Evolutionary Perspective on Food Review and Human Taste," https://www.ncbi.nlm.nih.gov/pmc/articles/PMC3680351/.

45. Sarah C. P. Williams, "Human Nose Can Detect a Trillion Smells," March 20, 2014, https://www.sciencemag.org/news/2014/03/human-nose-can-detect-trillion-smells.

46. Rakaia Kenney, Kayla Lemons, Weihong Lin, "What Makes Something Smell Good or Bad?" Curious Kids, June 1, 2020, https://umbc.edu/stories/what-makes-something-smell-good-or-bad/.

47. Sarah C. P. Williams, "Human Nose Can Detect a Trillion Smells," March 20, 2014, https://www.sciencemag.org/news/2014/03/human-nose-can-detect-trillion-smells.

48. Kenney, "What makes something smell good or bad?"

49. Dave Mosher, "Why Stuff Stinks: Secret Sniffed Out" LiveScience, September 21, 2007, https://www.livescience.com/4620-stuff-stinks-secret-sniffed.html.

50. "Nervous system — Taste," Science: Body & Mind, September 24, 2014, https://www.bbc.co.uk/science/humanbody/body/factfiles/taste/taste_animation.shtml.

51. Paul A.S. Breslin, "An Evolutionary Perspective on Food Review and Human Taste," https://www.ncbi.nlm.nih.gov/pmc/articles/PMC3680351/.

52. Dr. Christopher S. Baird, "Science Questions With Surprising Answers," December 14, 2012, https://www.wtamu.edu/cbaird/sq/2012/12/14/how-can-we-differentiate-so-many-different-foods-if-we-can-only-taste-four-flavors-on-our-tongue-sweet-bitter-sour-and-salty/.

53. Paul A.S. Breslin, "An Evolutionary Perspective on Food Review and Human Taste," https://www.ncbi.nlm.nih.gov/pmc/articles/PMC3680351/.

54. "Why Does Food Taste Good?" Medicareful Living, May 27, 2020, https://living.medicareful.com/why-does-food-taste-good.

55. Paul A.S. Breslin, "An Evolutionary Perspective on Food Review and Human Taste," https://www.ncbi.nlm.nih.gov/pmc/articles/PMC3680351/.

56. Ibid.

57. "Digestive System," https://my.clevelandclinic.org/health/articles/7041-the-structure-and-function-of-the-digestive-system.

58. Ibid.

59 "Saliva," ScienceDaily, https://www.sciencedaily.com/terms/saliva.htm.

60 Ibid.

61 "The Brain Stem," https://courses.lumenlearning.com/boundless-ap/chapter/the-brain-stem/.

62 Linda J. Vorvick, MD, "Peristalsis," MedlinePlus, https://medlineplus.gov/ency/anatomyvideos/000097.htm.

63 "Swallowing Exercises: Closure of the Larynx Exercises," Johns Hopkins Medicine, https://www.hopkinsmedicine.org/health/treatment-tests-and-therapies/swallowing-exercises-closure-of-the-larynx-exercises.

64 "The Digestive Process: How Does the Esophagus Work?" Stanford Children's Health, https://www.stanfordchildrens.org/en/topic/default?id=the-digestive-process-how-does-the-esophagus-work-134-195.

65 Linda J. Vorvick, MD, "Peristalsis," MedlinePlus, June 28, 2018, https://medlineplus.gov/ency/anatomyvideos/000097.htm.

66 "The Structure and Function of the Digestive System," September 13, 2018, https://my.clevelandclinic.org/health/articles/7041-the-structure-and-function-of-the-digestive-system.

67 Scott Frothingham, "How Big Is Your Stomach?" October 23, 2018, https://www.healthline.com/health/how-big-is-your-stomach.

68 "Digestive System," Quizlet on https://quizlet.com/11278288/digestive-system-flash-cards/.

69 Richard Bowen, "Absorption of Water and Electrolytes," May 2019, http://www.vivo.colostate.edu/hbooks/pathphys/digestion/smallgut/absorb_water.html.

70 "Digestive System," Quizlet, https://quizlet.com/11278288/digestive-system-flash-cards/.

71 "What Is a Bowel Transit Time Test?" https://www.webmd.com/a-to-z-guides/bowel-transit-time-test#1.

72 "The Small Intestine," http://oerpub.github.io/epubjs-demo-book/content/m46512.xhtml.

73 "The Structure and Function of the Digestive System," 09/13/2018, https://my.clevelandclinic.org/health/articles/7041-the-structure-and-function-of-the-digestive-system.

74 "Digestive System," Quizlet, https://quizlet.com/11278288/digestive-system-flash-cards/.

75 "What Is a Bowel Transit Time Test?" https://www.webmd.com/a-to-z-guides/bowel-transit-time-test#1.

76 "Accessory Organs in Digestion: The Liver, Pancreas, and Gallbladder," Chapter 13: The Digestive System, https://courses.lumenlearning.com/nemcc-ap/chapter/accessory-organs-in-digestion-the-liver-pancreas-and-gallbladder/.

77 "The Structure and Function of the Digestive System," September 13, 2018, https://my.clevelandclinic.org/health/articles/7041-the-structure-and-function-of-the-digestive-system.

78 Jenn Fusion, "What Are the Functions of a Liver Cell?" March 13, 2018, https://sciencing.com/functions-liver-cell-5106552.html.

79 "How many chemical reactions occur in the human body per second?" Niral, brainly.in/question/3560162.

80 Abi Badrick, "Absorption in the Large Intestine," June 3, 2018, https://teachmephysiology.com/gastrointestinal-system/large-intestine/absorption-large-intestine/.

81 "Feces," https://www.britannica.com/science/feces.

82 "What Is a Bowel Transit Time Test?" https://www.webmd.com/a-to-z-guides/bowel-transit-time-test#1.

83 Jess Speller, "Large Intestinal Motility," June 3, 2018, https://teachmephysiology.com/gastrointestinal-system/large-intestine/large-intestinal-motility/.

84 "Chapter 18: The Small and Large Intestines," BIO 140 — Human Biology I — Textbook, https://guides.hostos.cuny.edu/c.php?g=921955&p=6823613.

85 "The Structure and Function of the Digestive System," 09/13/2018, https://my.clevelandclinic.org/health/articles/7041-the-structure-and-function-of-the-digestive-system.

86 Jess Speller, "Large Intestinal Motility," June 3, 2018, https://teachmephysiology.com/gastrointestinal-system/large-intestine/large-intestinal-motility/.

87 "The Structure and Function of the Digestive System," 09/13/2018, https://my.clevelandclinic.org/health/articles/7041-the-structure-and-function-of-the-digestive-system.

88 Chaunie Brusie, "Is It Safe to Hold Your Pee?" February 6, 2017, https://www.healthline.com/health/holding-pee.

89 "How does the urinary system work?" InformedHealth.org, April 5, 2018, https://www.ncbi.nlm.nih.gov/books/NBK279384/.

90 M.L. Forsling, H. Montgomery, D. Halpin, R.J. Windle, D.F. Treacher, (May 1998). "Daily Patterns of Secretion of Neurohypophysial Hormones in Man: Effect of Age," National Library of Medicine, May 1988.

91 "Bedwetting," All for Kids Pediatric Clinic, https://www.afkpeds.org/our-services/sick-kids/bedwetting/.

92 "How Does the Urinary System Work?" National Library of Medicine, April 5, 2018, https://www.ncbi.nlm.nih.gov/books/NBK279384/.

93 "How Does the Urinary System Work?" National Library of Medicine, April 5, 2018, https://www.ncbi.nlm.nih.gov/books/NBK279384/.

94 "Is It Safe to Hold Your Pee?" by Chaunie Brusie — February 6, 2017, on https://www.healthline.com/health/holding-pee.

95 "Why Does Rotting Food Smell Bad?" Tufts University, December 16, 2014, https://phys.org/news/2014-12-food-bad.html.

96 "Can You Smell if Food Is Off?" ABC Health & Wellbeing, https://www.abc.net.au/news/health/2017-06-07/can-you-smell-if-food-is-off/8594238.

97 Amy Halloran, "When Odors Warn: What Does the Nose Know?" July 23, 2011, https://www.foodsafetynews.com/2011/07/when-odors-sound-a-warning-what-does-the-nose-know/.

98 Paul A.S. Breslin, "An Evolutionary Perspective on Food Review and Human Taste," https://www.ncbi.nlm.nih.gov/pmc/articles/PMC3680351/.

99 "Nervous system — Taste," BBC Science: Body & Mind, September 24, 2014, https://www.bbc.co.uk/science/humanbody/body/factfiles/taste/taste_animation.shtml.

100 "Blocked Intestine," Cedars Sinai, https://www.cedars-sinai.org/health-library/diseases-and-conditions/b/blocked-intestine.html.

101 Kristeen Moore, "Intestinal Obstruction," September 28, 2018, https://www.healthline.com/health/intestinal-obstruction.

102 Ibid.

103 Bryan Quoc Leon, "Umami and Kokumi — A Flavor Profile," August 23, 2017, on https://sciencemeetsfood.org/umami-kokumi-flavor-profile/.

104 "What Happens in My Body When I Vomit?" Science Focus, https://www.sciencefocus.com/the-human-body/what-happens-in-my-body-when-i-vomit.

105 "Brain Basics: Understanding Sleep," National Institute of Neurological Disorders and Stroke, https://www.ninds.nih.gov/disorders/patient-caregiver-education/understanding-sleep#top.

106 Eric Suni, "How Much Sleep Do We Really Need?" July 31, 2020, https://www.sleepfoundation.org/how-sleep-works/how-much-sleep-do-we-really-need.

107 Kristy Ramirez, "13 Baby Instincts That Kick in Immediately After Birth," Babygaga, April 6, 2017, https://www.babygaga.com/13-baby-instincts-that-kick-in-immediately-after-birth/.

108 Jessica Timmons, "How Long Does the Startle Reflex in Babies Last?" Healthline, January 12, 2018, https://www.healthline.com/health/parenting/startle-reflex-in-babies.

109 "Benefits Of Breastfeeding — Incredible Facts About Breastfeeding for Mommy & Baby," on http://www.breastfeedingrose.org/incredible-facts-about-breastfeeding-for-mommy-baby/.

110 Genevieve Howland, "Infant Reflexes: Are Those Weird Movements Normal?" December 5, 2019, https://mamanatural.com/infant-reflexes/.

111 Ibid.

112 Ibid.

113 "Innate Behavior of Animals," cK-12, https://read.activelylearn.com/#teacher/reader/authoring/preview/526033/notes.

114 "Newborn Reflexes," 3/8/21, https://www.healthychildren.org/English/ages-stages/baby/Pages/Newborn-Reflexes.aspx.

115 Ashley Marcin, "What Is Extrusion Reflex?" June 28, 2018, https://www.healthline.com/health/parenting/extrusion-reflex.

116 Jamie Eske, "What to Know about the Moro Reflex?" December 20, 2019, https://www.medicalnewstoday.com/articles/327370.

117 Ibid.

118 Genevieve Howland, "Infant Reflexes: Are Those Weird Movements Normal?" December 5, 2019, https://mamanatural.com/infant-reflexes/.

119 Ibid.

120 Jamie Eske, "What to Know about the Moro Reflex?" December 20, 2019, https://www.medicalnewstoday.com/articles/327370.

121 "Newborn Reflexes," March 8, 2021, https://www.healthychildren.org/English/ages-stages/baby/Pages/Newborn-Reflexes.aspx.

122 "Insight into How Infants Learn to Walk," Lancaster University, December 14, 2017, https://www.sciencedaily.com/releases/2017/12/171214101444.htm.

123 Genevieve Howland, "Infant Reflexes: Are Those Weird Movements Normal?" December 5, 2019, https://mamanatural.com/infant-reflexes/.

124 "Baby Reflexes: Hand-To-Mouth Is More Than It Seems!" Conscious Baby Blog, February 7, 2012, https://consciousbabyblog.wordpress.com/2012/02/07/baby-reflexes-hand-to-mouth-is-more-than-it-seems/.

125 Genevieve Howland, "Infant Reflexes: Are Those Weird Movements Normal?" December 5, 2019, https://mamanatural.com/infant-reflexes/.

126 Pierre Dukan, "What Muscles Are Involved in Walking?" Nutrition & Dietetics, https://www.sharecare.com/health/walking/what-muscles-involved-walking.

127 David Stauth, "One Step at a Time, Researchers Learning How Humans Walk," Oregon State University, January 17, 2014, https://phys.org/news/2014-01-humans.html.

128 "What Babies and Toddlers Know — and Learn," reviewed on January 9, 2019, https://www.whattoexpect.com/first-year/photo-gallery/what-babies-and-toddlers-know-and-learn.aspx#/slide-1.

129 "The Brain Consumes Half of a Child's Energy, and That Could Matter for Weight Gain," June 17, 2019, https://neurosciencenews.com/brain-development-food-1.

130 C. Claiborne Ray, "Brain Power," New York Times, September 1, 2008, https://www.nytimes.com/2008/09/02/science/02qna.html.

131 "Brain Development," https://www.firstthingsfirst.org/early-childhood-matters/brain-development/.

132 "Darwin's Greatest Challenge Tackled: The Mystery of Eye Evolution," November 1, 2004, https://www.sciencedaily.com/releases/2004/10/041030215105.htm.

133 Larger "Brain Anatomy and How the Brain Works," Health, Johns Hopkins Medicine, https://www.hopkinsmedicine.org/health/conditions-and-diseases/anatomy-of-the-brain.

134 Robert H. Spector, Clinical Methods: The History, Physical, and Laboratory Examinations, Chapter 58, The Pupils, https://www.ncbi.nlm.nih.gov/books/NBK381/.

135 "Rods and Cones," http://hyperphysics.phy-astr.gsu.edu/hbase/vision/rodcone.html.

136 E.R. Kandel, J.H. Schwartz, T.M. Jessell, Principles of Neural Science (New York: McGraw-Hill, 2000), p. 507–513.

137 Haruhisa Okawa, Alapakkam P. Sampath (2007). "Optimization of Single-Photon Response Transmission at the Rod-to-Rod Bipolar Synapse," American Physiology Society, August 2007, p. 279–286.

138 Ibid.

139 "Rods and Cones," http://hyperphysics.phy-astr.gsu.edu/hbase/vision/rodcone.html.

140 Ibid.

141 Brian A. Wandell, Stanford University, Foundations of Vision, "Chapter 3: The Photoreceptor Mosaic," https://foundationsofvision.stanford.edu/chapter-3-the-photoreceptor-mosaic/.

142 "How Does the Eye Determine Distance?" https://www.sharecare.com/health/eye-care/how-eye-determines-distance.

143 "How Do I See Depth?" http://scecinfo.usc.edu/geowall/ste.

144 https://www.vision3d.com/stereo.html.

145 Max Planck Society, "Brain Region Identified That Specializes in Close-up Exploration," August 16, 2019, https://medicalxpress.com/news/2019-08-brain-region-specializes-close-up-exploration.html.

146 AAHstaff, "What Is Directional Hearing? (And What Happens Without It)," 07/26/2019, https://advancedhearing.com/articles/what-directional-hearing-and-what-happens-without-it.

147 "Sound Source Localization," European Annals of Otorhinolaryngology, Head and Neck Diseases, Volume 135, Issue 4, August 2018, p. 259–264, https://www.sciencedirect.com/science/article/pii/S187972961830067X.

148 "Sound Localization," https://en.wikipedia.org/wiki/Sound_localization.

149 Joseph J. Volpe, Volpe's Neurology of the Newborn, "Neurological Examination" (Amsterdam: Elsevier Inc., 2018.)

150 Joseph J. Volpe, Volpe's Neurology of the Newborn, "Neurological Examination: Normal and Abnormal Features," 2018, https://www.sciencedirect.com/science/article/pii/B9780323428767000090.

151 Murphy Moroney, "Child Not Making Eye Contact? Here's What to Do, According to Experts," August 28, 2019, https://www.popsugar.com/family/How-to-Encourage-Eye-Contact-in-Babies-46524343.

152 "Why Do Dogs Make Eye Contact," https://wagwalking.com/behavior/why-do-dogs-make-eye-contact.

153 Colleen de Bellefonds, "Here's a Scientific Reason to Keep Gazing Into Your Baby's Eyes (Just in Case You Needed Another Excuse!)". December 1, 2017, https://www.whattoexpect.com/news/first-year/.

154 Michael D. Coe, "The Language Within Us" The New York Times, February 27, 1994, https://archive.nytimes.com/www.nytimes.com/books/98/12/06/specials/pinker-instinct.html, a review of The Language Instinct by Steven Pinker (New York: William Morrow & Company, 1994).

155 Steven Pinker, The Language Instinct (New York: William Morrow & Company, 1994).

156 Steven Pinker, "Language Is a Human Instinct," https://www.edge.org/conversation/steven_pinker-chapter-13-language-is-a-human-instinct.

157 Ibid.

158 Michael D. Coe, "The Language Within Us," The New York Times, February 27, 1994, https://archive.nytimes.com/www.nytimes.com/books/98/12/06/specials/pinker-instinct.html.

159 Kristy Ramirez, "13 Baby Instincts That Kick in Immediately After Birth," Babygaga, April 6, 2017, https://www.babygaga.com/13-baby-instincts-that-kick-in-immediately-after-birth/.

160 Erica Hersh, "How Many Times Do You Blink in a Day?" September 24, 2020, https://www.healthline.com/health/how-many-times-do-you-blink-a-day.

161 Ben Mauk, "Why Do We Blink?" October 24, 2012, https://www.livescience.com/32189-why-do-we-blink.html.

162 Erica Hersh, "How Many Times Do You Blink in a Day?" September 24, 2020, https://www.healthline.com/health/how-many-times-do-you-blink-a-day.

163 Ben Mauk, "Why Do We Blink?" October 24, 2012, https://www.livescience.com/32189-why-do-we-blink.html.

164 Kristy Ramirez, "13 Baby Instincts That Kick in Immediately After Birth," Babygaga, April 6, 2017, https://www.babygaga.com/13-baby-instincts-that-kick-in-immediately-after-birth/.

165 Ibid.

166 Bryan Johnson, "Top 10 Human Reflexes and Natural Instincts," January 28,2012, https://listverse.com/2012/01/28/top-10-human-reflexes-and-natural-instincts/.

167 "Babies Can Skydive!" September 9, 2016, http://www.smallacorn.co.uk/fff/week-37/.

168 Ibid.

169 Kristy Ramirez, "13 Baby Instincts That Kick in Immediately After Birth," Babygaga, April 6, 2017, https://www.babygaga.com/13-baby-instincts-that-kick-in-immediately-after-birth/.

170 Ibid./

171 "What Do Your Baby's Cries Mean?" by Catherine Donaldson-Evans, reviewed on April 15, 2019, on https://www.whattoexpect.com/first-year/week-10/decoding-cries.aspx. Also look at "Baby Crying Sounds – What Do Different Cries Mean?" on https://www.petitjourney.com.au/understand-the-different-cries-of-your-baby/.

172 Ibid.

173 Dr. Paul Brand and Philip Yancey, The Gift of Pain (Grand Rapids, MI: Zondervan, 1997).

174 "Pain," Anatomy and Physiology I, Module 14: Sensory Systems, https://courses.lumenlearning.com/austincc-ap1/chapter/pain/.

175 "Pain and How You Sense It," https://www.mydr.com.au/pain-and-how-you-sense-it/.

176 Joseph J. Volpe, "Neurological Examination," Volpe's Neurology of the Newborn (Sixth Edition), 2018, https://www.sciencedirect.com/science/article/pii/B9780323428767000090.

177 "Benefits Of Breastfeeding – Incredible Facts About Breastfeeding for Mommy & Baby," http://www.breastfeedingrose.org/incredible-facts-about-breastfeeding-for-mommy-baby/.

178 "Why Do Dogs Make Eye Contact," https://wagwalking.com/behavior/why-do-dogs-make-eye-contact.

179 James Rowland Angell, "Psychology, Chapter 16: The Important Human Instincts," https://brocku.ca/MeadProject/Angell/Angell_1906/Angell_1906_p.html.

180 "Benefits Of Breastfeeding – Incredible Facts About Breastfeeding for Mommy & Baby," http://www.breastfeedingrose.org/incredible-facts-about-breastfeeding-for-mommy-baby/.

181 Melinda Ayre, "The Secret Behind Your Mama Bear Instinct," Honey Parenting, https://honey.nine.com.au/parenting/science-discovers-how-mothers-protective-instinct-works-when-children-threatened/8105c445-26d5-4b61-8890-720732a97de8.

182 Sarah Gibbens, "Is Maternal Instinct Only for Moms? Here's the Science," National Geographic, https://www.nationalgeographic.com/science/article/mothers-day-2018-maternal-instinct-oxytocin-babies-science.

183 Melinda Ayre, "The Secret Behind Your Mama Bear Instinct," Honey Parenting, https://honey.nine.com.au/parenting/science-discovers-how-mothers-protective-instinct-works-when-children-threatened/8105c445-26d5-4b61-8890-720732a97de8.

184 Sharon Palmer, RD, "Taking Control of Hunger — Lessons on Calming Appetite and Managing Weight," Today's Dietitian, April 2009, https://www.todaysdietitian.com/newarchives/040609p28.shtml.

185 David Reuben, M.D., Everything You Always Wanted to Know about Nutrition (New York: Maleri Corp., 1978).

186 Ryan Raman, MS, RD, "Feeling Hungry After Eating: Why It Happens and What to Do," May 28, 2020, https://www.healthline.com/nutrition/feeling-hungry-after-eating#causes-solutions.

187 Joseph Proietto, Professor of Medicine, "Chemical Messengers: How Hormones Make Us Feel Hungry and Full," September 25, 2015, https://theconversation.com/chemical-messengers-how-hormones-make-us-feel-hungry-and-full-35545.

188 Melinda Beck, "Why That Big Meal You Just Ate Made You Hungry," April 14, 2009, The Wall Street Journal, https://www.wsj.com/articles/SB123966898930315491.

189 Ryan Raman, MS, RD, "Feeling Hungry After Eating: Why It Happens and What to Do," May 28, 2020, https://www.healthline.com/nutrition/feeling-hungry-after-eating#causes-solutions.

190 Here's Why You Don't Feel Full After Eating Junk Food by Katy Severson, updated March 11, 2019 on www.huffpost.com.

191 "Why Are Americans Obese?" https://www.publichealth.org/public-awareness/obesity/.

192 Barbara Kollmeyer, "The U.S. Is the Most Obese Nation in the World, Just Ahead of Mexico," May 29, 2017, https://www.marketwatch.com/story/the-us-is-the-most-obese-nation-in-the-world-just-ahead-of-mexico-2017-05-19.

193 Edmund Custers, "Reflex Action and Reflex Arc: What Happens When You Accidentally Touch a Hot Pot," January 31, 2019, https://owlcation.com/stem/Here-is-what-happens-when-you-accidentally-touch-a-hot-pot.

194 "The Somatic Nervous System," Pressbooks, http://neuroscience.openetext.utoronto.ca/chapter/anatomy-physiology-the-somatic-nervous-system/.

195 Celena Derderian, "Physiology, Withdrawal Response," September 13, 2021, https://www.ncbi.nlm.nih.gov/books/NBK544292/.

196 Edmund Custers, "Reflex Action and Reflex Arc: What Happens When You Accidentally Touch a Hot Pot," January 31, 2019, https://owlcation.com/stem/Here-is-what-happens-when-you-accidentally-touch-a-hot-pot.

197 "Hand on a Hot Stove," https://www.urmc.rochester.edu/medialibraries/urmcmedia/life-sciences-learning-center/documents/2013-14neuroscience/handonhotstove-092513-student.pdf.

198 Sarah C. P. Williams, "Human Nose Can Detect a Trillion Smells," March 20, 2014, https://www.sciencemag.org/news/2014/03/human-nose-can-detect-trillion-smells.

199 Dave Mosher, "Why Stuff Stinks: Secret Sniffed Out," September 21, 2007, Home News, Live Science, https://www.livescience.com/4620-stuff-stinks-secret-sniffed.html.

200 Alex Charfen, "Natural Thirst: The Instinct You Didn't Know You Lost," March 26, 2017, Huffington Post, https://www.huffpost.com/entry/natural-thirst-the-instin_b_9547500.

201 "How Do Our Noses 'Adjust' to Bad Smells?" June 5, 2021, https://www.mentalfloss.com/article/53526/how-do-our-noly.

202 "The Black Death: The Plague, 1331–1770," Hardin Library for the Health Sciences, http://hosted.lib.uiowa.edu/histmed/plague/.

203 Ibid.

204 Bryan Johnson, "Top 10 Human Reflexes and Natural Instincts," January 28, 2012, https://listverse.com/2012/01/28/top-10-human-reflexes-and-natural-instincts/.

205 "The Brain Stem" in Lumen, Boundless Anatomy and Physiology, https://courses.lumenlearning.com/boundless-ap/chapter/the-brain-stem/.

206 Bryan Johnson, "Top 10 Human Reflexes and Natural Instincts," January 28, 2012, https://listverse.com/2012/01/28/top-10-human-reflexes-and-natural-instincts/.

207 David H. Nguyen, Ph.D., "How the Immune Response Contributes to Homeostasis," April 27, 2018, https://sciencing.com/immune-response-contributes-homeostasis-20915.html.

208 Ibid.

209 "Homeostasis," Bitesize, https://www.bbc.co.uk/bitesize/guides/zqgfv4j/revision/3.

210 "Maintaining Stable Body Conditions," Bitesize, https://www.bbc.co.uk/bitesize/guides/znc6fg8/revision/4.

211 "Why Do I Shiver When I'm Cold?" Children's Museum Indianapolis, https://www.childrensmuseum.org/blog/why-do-i-shiver-when-i'm-cold.

212 James Roland, "What You Should Know About Shivering," September 18, 2018, https://www.healthline.com/health/shivering.

213 "Homeostasis," Bitesize, https://www.bbc.co.uk/bitesize/guides/zqgfv4j/revision/4.

214 Stephen J. Praetorius, "9 Facts You Didn't Know About Sweat," December 8, 2015, https://www.gq.com/story/9-facts-you-didnt-know-about-sweat.

215 "Sweating," https://medlineplus.gov/ency/anatomyvideos/000127.htm.

216 Stephen J. Praetorius, "9 Facts You Didn't Know About Sweat," December 8, 2015, https://www.gq.com/story/9-facts-you-didnt-know-about-sweat.

217 Ibid.

218 "Homeostasis," Bitesize, https://www.bbc.co.uk/bitesize/guides/zqgfv4j/revision/3.

219 "Hyperglycemia in Diabetes," https://www.mayoclinic.org/diseases-conditions/hyperglycemia/symptoms-causes/syc-20373631.

220 "Converting Carbohydrates to Triglycerides," https://www.ncsf.org/pdf/ceu/converting_carbohydrates_to_triglycerides.pdf; "What is Glucagon?" Hormone Health Network, https://www.hormone.org/your-health-and-hormones/glands-and-hormones-a-to-z/hormones/glucagon; "Glucagon," January 15, 2019, https://www.diabetes.co.uk/body/glucagon.html.

221 "What is Glucagon?" Hormone Health Network, https://www.hormone.org/your-health-and-hormones/glands-and-hormones-a-to-z/hormones/glucagon.

222 Macarena Pozo and Marc Claret, "Hypothalamic Control of Systemic Glucose Homeostasis: The Pancreas Connection," Trends in Endocrinology & Metabolism, https://www.cell.com/trends/endocrinology-metabolism/comments/S1043-2760(18)30106-1.

223 Dawn Skelton, "Explainer: Why Does Our Balance Get Worse as We Grow Older?" The Conversation, October 8, 2015, https://theconversation.com/explainer-why-does-our-balance-get-worse-as-we-grow-older-48197.

224 "Our Sense of Balance," Iowa Ear Center, https://www.iowaearcenter.com/our-sense-of-balance.

225 Dawn Skelton, "Explainer: Why Does Our Balance Get Worse as We Grow Older?" The Conversation, October 8, 2015, https://theconversation.com/explainer-why-does-our-balance-get-worse-as-we-grow-older-48197.

226 Timothy C. Hain, MD, "Otoliths," March 9, 2021, https://dizziness-and-balance.com/disorders/bppv/otoliths.html.

227 "Our Sense of Balance," The Balance Clinic, https://www.iowaearcenter.com/our-sense-of-balance.

228 Robert Roy Britt, "Does Blue Light From Screens Really Ruin Sleep? " February 13, 2020, https://elemental.medium.com/does-blue-light-from-screens-really-ruin-sleep-ae5e758c453e.

229 Ibid.

230 "Fingertips to Hair Follicles: Why 'Touch' Triggers Pleasure and Pain," NPR News, February 3, 2015, https://www.wbur.org/npr/383426166/fingertips-to-hair-follicles-why-touch-causes-pleasure-and-pain.

231 Dr. Erica Wang, "Top 5 Fever Myths and Facts," Texas Children's Hospital, https://www.texaschildrens.org/blog/2016/11/top-5-fever-myths-and-facts.

232 David H. Nguyen, Ph.D., "How the Immune Response Contributes to Homeostasis," April 27, 2018, https://sciencing.com/immune-response-contributes-homeostasis-20915.html.

233 Ibid.

234 Ibid.

235 Ibid.

236 Kerry Laing, et al., "Immune responses to viruses," in British Society for Immunology, https://www.immunology.org/ public-information/bitesized-immunology/pathogens-and- disease/immune-responses-viruses.

237 David H. Nguyen, Ph.D., "How the Immune Response Contributes to Homeostasis," April 27, 2018, https://sciencing.com/immune- response-contributes-homeostasis-20915.html.

238 "How Influenza (Flu) Vaccines Are Made," Centers for Disease Control, August 31, 2021, https://www.cdc.gov/flu/prevent/how-fluvaccine-made.htm.

239 "Understanding How COVID-19 Vaccines Work," Centers for Disease Control and Prevention, updated December 14, 2021, www.cdc.gov/coronavirus/2019-ncov/vaccines/different-vaccines/how-they-work.html.

240 "Functions of the Liver," The Hepatitus C Trust, http://hepctrust.org.uk/information/liver/functions-liver.

241 Jenn Fusion, "What Are the Functions of a Liver Cell?" March 13, 2018, https://sciencing.com/functions-liver-cell-5106552.html.

242 Ibid.

243 "Functions of the Liver," The Hepatitus C Trust, http://hepctrust.org.uk/information/liver/functions-liver.

244 "How Brain Can Distinguish Good from Bad Smells," ScienceDaily, Max Planck Institute for Chemical Ecology, December 16, 2014, https://www.sciencedaily.com/releases/2014/12/141216100519.htm.

245 "Fears and Phobias," TeensHealth, https://kidshealth.org/en/teens/phobias.html.

246 Ibid.

247 "How Brain Can Distinguish Good from Bad Smells," ScienceDaily, Max Planck Institute for Chemical Ecology, December 16, 2014, https://www.sciencedaily.com/releases/2014/12/141216100519.htm.

248 Clara Moskowitz, "Why We Fear Snakes," March 3, 2008, https://www.livescience.com/2348-fear-snakes.html.

249 K. Pacak, R. McCarty, "Acute Stress Response: Experimental," Encyclopedia of Stress (Second Edition), 2007, https://www.sciencedirect.com/topics/nursing-and-health-professions/suffocation.

250 Charles Q. Choi, "Carbon Dioxide Triggers Primordial Fear of Suffocation," November 25, 2009, https://www.livescience.com/5910-carbon-dioxide-triggers-primordial-fear-suffocation.html.

251 "Instinctive Drowning Response: Know More," January 20, 2016, https://www.thebeachcompany.in/blogs/be-beachy/105947655-instinctive-drowning-response-know-more.

252 John R. Fletemeyer, "What Really Happens When Someone Drowns?" Aquatics International, April 4, 2017, https://www.aquaticsintl.com/lifeguards/what-really-happens-when-someone-drowns_o.

253 An excellent podcast called, "I Don't Have Enough Faith to Be an Atheist" by Dr. Frank Turek.

254 "Self-preservation Is the First Law of Nature," The Free Dictionary by Farlex, https://idioms.thefreedictionary.com/self-preservation+is+the+first+law+of+nature.

255 Casey Luskin, "No, Scientists in Darwin's Day Did Not Grasp the Complexity of the Cell; Not Even Close," June 6, 2013, https://evolutionnews.org/2013/06/did_scientists_/.

256 Robert Shapiro, Origins (Bantam Books, 1987), p. 57–58.

257 "Why Do Fishes Die When They're Removed from Water?" UCSB ScienceLine, http://scienceline.ucsb.edu/getkey.php?key=4461.

258 "Why Do Fish Flop When They Are Out of Water?" CastCountryTeam, https://castcountry.com/why-do-fish-flop-when-they-are-out-of-water.

259 Bruce Hecker, "How do Whales and Dolphins Sleep Without Drowning?" Scientific American, February 2, 1998, https://www.scientificamerican.com/article/how-do-whales-and-dolphin/.

260 David E. Leib, "Thirst," https://www.ncbi.nlm.nih.gov/pmc/articles/PMC5957508/.

261 Sonia Madaan, "How Can Animals Drink Dirty Water?" https://eartheclipse.com/animals/how-can-animals-drink-dirty-water.html.

262 Philip J. Ryan, "The Neurocircuitry of Fluid Satiation," in Physiological Reports, https://www.ncbi.nlm.nih.gov/pmc/articles/PMC6014472/.

263 Sonia Madaan, "How Can Animals Drink Dirty Water?" https://eartheclipse.com/animals/how-can-animals-drink-dirty-water.html.

264 "What We Can Learn from a Giraffe's Neck," https://www.pfizer.com/foundations-science/what-we-can-learn-giraffe's-neck.

265 Yiming Chen and Zachary A. Knight, "Making Sense of the Sensory Regulation of Hunger Neurons," National Library of Medicine, https://www.ncbi.nlm.nih.gov/pmc/articles/PMC4899083/.

266 Elaine Magee, "Your 'Hunger Hormones'; How They Affect Your Appetite and Your Weight," https://www.webmd.com/diet/features/your-hunger-hormones.

267 Yiming Chen and Zachary A. Knight, "Making Sense of the Sensory Regulation of Hunger Neurons," National Library of Medicine, https://www.ncbi.nlm.nih.gov/pmc/articles/PMC4899083/.

268 Tetsuya Miyamoto, Geraldine Wright, and Hubert Amrein, "Nutrient sensors," May 6, 2013, https://www.ncbi.nlm.nih.gov/pmc/articles/PMC4332773/.

269 Ashish, "How Can Wild Animals Drink Water from Dirty Ponds and Lakes and Not Get Sick?" Science ABC, https://www.scienceabc.com/eyeopeners/how-can-wild-animals-drink-water-from-dirty-ponds-and-lakes-and-not-get-sick.html.

270 G.M. Klump, "Localizing Signal Sources," ScienceDirect, https://www.sciencedirect.com/topics/medicine-and-dentistry/directional-hearing.

271 Ibid.

272 Marc Bekoff, "Animal Instincts: Not What You Think They Are," March 8, 2011, https://greatergood.berkeley.edu/article/item/animal_instincts.

273 "Amazing Facts About Bats," The Nature Conservancy, August 12, 2019, https://www.nature.org/en-us/about-us/where-we-work/united-states/arizona/stories-in-arizona/top-10-bat-facts/.

274 "Flight, Food and Echolocation," Bat Conservation Trust, https://www.bats.org.uk/about-bats/flight-food-and-echolocation.

275 "How Do Birds Know When You Fill the Feeder?" August 31, 2020, https://www.whatbirdsareinmybackyard.com/2020/08/how-do-birds-know-when-you-fill-feeder.html.

276 Nick Stockton, "What's Up with That: Birds Bob Their Heads When They Walk," 1-20-2015, https://www.wired.com/2015/01/whats-birds-bob-heads-walk/.

277 Ibid.

278 Joan Morris, "Do Animals Instinctively Know What Not to Eat?" https://www.mercurynews.com/2018/01/30/do-animals-instinctively-know-what-not-to-eat/.

279 Ibid.

280 Paul A.S. Breslin, "An Evolutionary Perspective on Food Review and Human Taste," https://www.ncbi.nlm.nih.gov/pmc/articles/PMC3680351/.

281 Ibid.

282 "How Brain Can Distinguish Good from Bad Smells," ScienceDaily, December 16, 2014, https://www.sciencedaily.com/releases/2014/12/141216100519.htm.

283 Ibid.

284 Ibid.

285 Ibid.

286 Rakaia Kenney, "What Makes Something Smell Good or Bad?" The Conversation, Curious Kids, June 1, 2020, https://theconversation.com/what-makes-something-smell-good-or-bad-136929.

287 Dan Moore, "Why Can Animals Eat Raw Meat? 7 Amazing Facts (Explained)," Animal How, https://animalhow.com/animals-raw-meat/.

288 "Sharks of Hawaii" video on PBS, viewed on April 22, 2021.

289 Paul A. Bartz, "Catching Bigger, Stronger and Faster Prey," Creation Moments, August 2, 2021.

290 "So, What Do Jellyfish Eat and How Do They Eat It?" by AZ Animals Staff on https://a-z-animals.com/blog/so-what-do-jellyfish-eat-and-how-do-they-eat-it/

291 J. David Sweatt, "Non-Associative Learning and Memory," Mechanisms of Memory (Second Edition), 2010, https://www.sciencedirect.com/topics/biochemistry-genetics-and-molecular-biology/tritonia.

292 Jennie Eilerts, "How Does the Digestive System Work in a Cow: Understanding the Ruminant Digestive System," June 13, 2019, https://proearthanimalhealth.com/how-does-the-digestive-system-work-in-a-cow-understanding-the-ruminant-digestive-system/.

293 "Rumen Physiology and Rumination," VIVO Pathophysiology, http://www.vivo.colostate.edu/hbooks/pathphys/digestion/herbivores/rumination.html.

294 Jennie Eilerts, "How Does the Digestive System Work in a Cow: Understanding the Ruminant Digestive System," June 13, 2019, https://proearthanimalhealth.com/how-does-the-digestive-system-work-in-a-cow-understanding-the-ruminant-digestive-system/.

295 "Rumen Physiology and Rumination," VIVO Pathophysiology, http://www.vivo.colostate.edu/hbooks/pathphys/digestion/herbivores/rumination.html.

296 "Why Do Humans and Many Other Animals Sleep?" National Library of Medicine, https://www.ncbi.nlm.nih.gov/books/NBK11108/.

297 Sara Harrison, "What Octopus Dreams Tell Us About the Evolution of Sleep," Wired, April 16, 2021, https://www.wired.com/story/what-octopus-dreams-tell-us-about-the-evolution-of-sleep/.

298 Andrew S. Freiberg, "Why We Sleep: A Hypothesis for an Ultimate or Evolutionary Origin for Sleep and Other Physiological Rhythms," National Library of Medicine, March 30, 2020, https://www.ncbi.nlm.nih.gov/pmc/articles/PMC7120898/.

299 Veronique Greenwood, "Sleep Evolved Before Brains. Hydras Are Living Proof," May 18, 2021, www.quantamagazine.org/.

300 "Why Do Humans and Many Other Animals Sleep?" National Library of Medicine, https://www.ncbi.nlm.nih.gov/books/NBK11108/.

301 Sara Harrison, "What Octopus Dreams Tell Us About the Evolution of Sleep," Wired, April 16, 2021, https://www.wired.com/story/what-octopus-dreams-tell-us-about-the-evolution-of-sleep/.

302 "Brain Basics: Understanding Sleep," National Institute of Neurological Disorders and Stroke, https://www.ninds.nih.gov/Disorders/Patient-Caregiver-Education/Understanding-Sleep#7.

303 D. Purves, G.J. Augustine, D. Fitzpatrick, et al. "Why Do Humans and Many Other Animals Sleep?" National Library of Medicine, https://www.ncbi.nlm.nih.gov/books/NBK11108/.

304 Ibid.

305 "Sharks of Hawaii" video on PBS, viewed on April 22, 2021.

306 "Do All Animals Need Sleep?" https://www.bbc.com/future/article/20130927-do-all-animals-need-sleep.

307 D. Purves, G.J. Augustine, D. Fitzpatrick, et al. "Why Do Humans and Many Other Animals Sleep?" National Library of Medicine, https://www.ncbi.nlm.nih.gov/books/NBK11108/.

308 "Elephant seals drift off to sleep while diving far below the ocean surface" by University of California - Santa Cruz in https://phys.org/news/2023-04-elephant-drift-ocean-surface.

309 Salama Yusuf, "Why Don't Birds Fall Off Branches When They Sleep?" November 13, 2021, https://www.scienceabc.com/nature/animals/why-dont-birds-fall-off-branches-when-they-sleep.html.

310 "Do Birds Sleep?" The Cornell Lab, https://www.birds.cornell.edu/k12/do-birds-sleep/.

311 Merrit Kennedy, "Scientists Pinpoint How a Flamingo Balances on One Leg," The Two-Way, May 25, 2017, https://www.npr.org/sections/thetwo-way/2017/05/25/530046238/scientists-pinpoint-how-a-flamingo-balances-on-one-leg.

312 "Do Birds Sleep?" The Cornell Lab, https://www.birds.cornell.edu/k12/do-birds-sleep/.

313 Amy Tikkanen, "Why Do Horses Sleep Standing Up?" https://www.britannica.com/story/why-do-horses-sleep-standing-up.

314 Ibid.

315 Marion Sipe, "What Is Innate and Learned Animal Behavior?" April 24, 2018, https://sciencing.com/innate-learned-animal-behavior-6668264.html.

316 "Geese" in Today's Creation Moment: April 20, 2021, https://creationmoments.com/sermons/geese/

317 Kateanswers, "Is Walking an Instinctive Behavior or a Learned Behavior?" eNotes Science, https://www.enotes.com/homework-help/walking-an-instinctive-behavior-learned-behavior-670784.

318 Marion Sipe, "What Is Innate and Learned Animal Behavior?" April 24, 2018, https://sciencing.com/innate-learned-animal-behavior-6668264.html.

319 D. Robert, "Directional Hearing Using Mechanically Coupled Pressure Receivers," Science Direct, https://www.sciencedirect.com/topics/biochemistry-genetics-and-molecular-biology/directional-hearing.

320 "How Do Fish Swim in Schools? Scientists Say It's in Their Genes," LiveScience, 09/18/2013, https://www.huffpost.com/entry/fish-schools-genes_n_3947303.

321 Matt Wedel, "Birds Have Balance Organs in Their Butts. Why Is No-one Talking about This?!" January 23, 2019, Sauropod Vertebra Picture of the Week, https://svpow.com/2019/01/23/birds-have-balance-organs-in-their-butts-why-is-no-one-talking-about-this/.

322 Darío Urbina-Meléndez, "A Physical Model Suggests That Hip-Localized Balance Sense in Birds Improves State Estimation in Perching: Implications for Bipedal Robots," National Library of Medicine, April 4, 2018, https://www.ncbi.nlm.nih.gov/pmc/articles/PMC7806032/.

323 Kathryn E. Stanchak, et al., "The Balance Hypothesis for the Avian Lumbosacral Organ and an Exploration of its Morphological Variation," bioRxiv, January 4, 2020, by Cold Spring Harbor Laboratory, https://www.biorxiv.org/content/10.1101/2020.04.01.020982v1.

324 Meg Michelle, "How Do Animals Communicate?" April 24, 2017, https://sciencing.com/animals-communicate-4566453.html.

325 "Instinct," https://science.jrank.org/pages/3611/Instinct.html.

326 "Innate behaviors," https://www.khanacademy.org/science/ap-biology/ecology-ap/responses-to-the-environment/a/innate-behaviors.

327 "Instinct," https://science.jrank.org/pages/3611/Instinct.html.

328 Michael Denton, Evolution: A Theory in Crisis (Chevy Chase, MD: Adler & Adler Publishers, Inc., 1986) p. 202.

329 Marion Sipe, "What Is Innate and Learned Animal Behavior?" April 24, 2018, https://sciencing.com/innate-learned-animal-behavior-6668264.html.

330 Marion Sipe, "What Is Innate and Learned Animal Behavior?" April 24, 2018, https://sciencing.com/innate-learned-animal-behavior-6668264.html.

331 Dean Mobbs et al., "The Ecology of Human Fear: Survival Optimization and the Nervous System," Frontiers in Neuroscience, March 18, 2015, https://www.ncbi.nlm.nih.gov/pmc/articles/PMC4364301/.

332 Dean Mobbs et al., "Neural Activity Associated with Monitoring the Oscillating Threat Value of a Tarantula," National Library of Medicine, https://www.ncbi.nlm.nih.gov/pmc/articles/PMC2996708/.

333 Dr. Frank Turek, weekly podcast "I Don't Have Enough Faith to Be an Atheist" on CrossExamined.org.

334 "Monkeys Hard-wired to Fear Snakes, Study Finds," The Australian, October 29, 2013, https://www.theaustralian.com.au/news/health-science/monkeys-hard-wired-to-fear-snakes-study-finds/news-story/712d07a319dd86802755b4ed123fb692.

335 "Instinct," https://science.jrank.org/pages/3611/Instinct.html.

336 Adam D. Douglass, Sebastian Kraves, Karl Deisseroth, Alexander F. Schier, and Florian Engert, "Escape Behavior Elicited by Single, Channelrhodopsin-2-Evoked Spikes in Zebrafish Somatosensory Neurons," National Library of Medicine, https://www.ncbi.nlm.nih.gov/pmc/articles/PMC2891506/.

337 Adam C. Roberts, Kaycey C. Pearce, Ronny C. Choe, Joseph B. Alzagatiti, Anthony K. Yeung, Brent R. Bill, David L. Glanzman, "Long-term Habituation of the C-start Escape Response in Zebrafish Larvae," National Library of Medicine, https://www.ncbi.nlm.nih.gov/pmc/articles/PMC5031492.

338 R.C. Eaton, R.K. Lee, M.B. Foreman, "The Mauthner Cell and Other Identified Neurons of the Brainstem Escape Network of Fish," National Library of Medicine, https://pubmed.ncbi.nlm.nih.gov/11163687/.

339 J. David Sweatt, "Non-Associative Learning and Memory," Mechanisms of Memory (Second Edition), 2010, https://www.sciencedirect.com/topics/biochemistry-genetics-and-molecular-biology/tritonia.

340 Paul S. Katz, "Tritonia Swim Network," Scholarpedia, http://www.scholarpedia.org/article/Tritonia_swim_network.

341 Katie Sokolowski, Joshua G. Corbin, "Wired for Behaviors: From Development to Function of Innate Limbic System Circuitry," National Library of Medicine, https://www.ncbi.nlm.nih.gov/pmc/articles/PMC3337482.

342 Charles D. Drewes, "Escape Reflexes in Earthworms and Other Annelids," Neural Mechanisms of Startle Behavior (Boston, MA: Springer, 1984), p. 43–91, https://link.springer.com/chapter/10.1007%2F978-1-4899-2286-1_3.

343 W. Johnson, P.D. Soden, E.R. Trueman, "A Study in Jet Propulsion: An Analysis of the Motion of the Squid, Loligo Vulgaris," Journal of Experimental Biology (1972), https://journals.biologists.com/jeb/article/56/1/155/21746/A-Study-in-Jet-Propulsion-An-Analysis-of-the.

344 T.S. Otis and W.F. Gilly, "Jet-propelled Escape in the Squid Loligo Opalescens: Concerted Control by Giant and Non-giant Motor Axon Pathways," Proceedings of the National Academy of Sciences of the USA, April 1, 1990, https://www.pnas.org/content/87/8/2911.

345 Franklin B. Krasne, "Excitation and Habituation of the Crayfish Escape Reflex: The Depolarizing Response in Lateral Giant Fibres of the Isolated Abdomen," Journal of Neuroscience, May 14, 1968, https://journals.biologists.com/jeb/article/50/1/29/21494/Excitation-and-Habituation-of-the-Crayfish-Escape.

346 D. Booth, B. Marie, P. Domenici, J.M. Blagburn, J.P. Bacon, "Transcriptional Control of Behavior: Engrailed Knock-Out Changes Cockroach Escape Trajectories," Journal of Neuroscience, https://www.ncbi.nlm.nih.gov/pmc/articles/PMC2744400/.

347 Garreth, "18 Animals that Play Dead (A to Z List & Pictures)," Fauna Facts, January 1, 2022, https://faunafacts.com/animals/examples-of-animals-that-play-dead/.

348 Patricia Greene, "16 Animals That Play Dead (Why They Do It & Pictures," Wildlife Informer, https://wildlifeinformer.com/animals-that-play-dead/.

349 Garreth, "18 Animals that Play Dead (A to Z List & Pictures)," Fauna Facts, January 1, 2022, https://faunafacts.com/animals/examples-of-animals-that-play-dead/.

350 Ibid.

351 Sarah Cairoli, "What Is the Prey in an Ecosystem?" April 25, 2017, https://sciencing.com/prey-ecosystem-4488.html.

352 "Playing Possum: 9 Animals That Play Dead to Survive," AZ Animals Staff, August 27, 2021, https://a-z-animals.com/blog/playing-possum-9-animals-that-play-dead-to-survive/.

353 Suzan, "Why Do Squirrels Have Jerky Movements?" Squirrel Enthusiast, https://squirrelenthusiast.com/why-do-squirrels-have-jerky-movements/.

354 Donald H. Owings, "Tonic Communication in the Antipredator Behavior of Ground Squirrels," ScienceDirect, https://www.sciencedirect.com/topics/agricultural-and-biological-sciences/self-preservation.

355 "In the Field: An Interview with Roger Hanlon," Journal of Experimental Biology, https://journals.biologists.com/jeb/article/224/24/jeb243831/273772?utm_campaign=JEBSnippet.

356 Mary Bates, "How Do Chameleons Change Colors?" Wired, April 11, 2014, https://www.wired.com/2014/04/how-do-chameleons-change-colors/.

357 Rebecca Bragg, "What Is the Process of a Chameleon Camouflaging Themselves?" Pets on Mom.com, animals.mom.com/process-chameleon-camouflaging-themselves-4521.html.

358 Dean Mobbs et al., "The Ecology of Human Fear: Survival Optimization and the Nervous System," Frontiers in Neuroscience, March 189, 2015, https://www.ncbi.nlm.nih.gov/pmc/articles/PMC4364301/.

359 Ibid.

360 Ibid.

361 "How Do Snakes Protect Themselves?" https://oddlycutepets.com/how-do-snakes-protect-themselves/.

362 "Snake Behavior and Life History — Defensive Strategies," A Guide to the Amphibians and Reptiles of California, http://www.californiaherps.com/behavior/snakelifehistorydefense.html.

363 "Snake Behavior and Life History — Defensive Strategies," A Guide to the Amphibians and Reptiles of California, http://www.californiaherps.com/behavior/snakelifehistorydefense.html.

364 "How Do Snakes Protect Themselves?" https://oddlycutepets.com/how-do-snakes-protect-themselves/.

365 Patrick Pester, "Humans Are Practically Defenseless. Why Don't Wild Animals Attack Us More?" Life's Little Mysteries, July 12, 2021, https://www.livescience.com/why-predators-dont-attack-humans.html.

366 Max Carol, "Animals Are More Afraid of Humans Than of Bears, Wolves, and Dogs," Treehugger, October 11, 2018, https://www.treehugger.com/animals-are-more-afraid-humans-bears-wolves-and-dogs-4858177.

367 "What Is the Meaning of Animal Instinct?" Expert Answers, enotes, https://www.enotes.com/homework-help/what-meaning-animal-instinct-150131.

368 "Instinct," https://science.jrank.org/pages/3611/Instinct.html.

369 Mark S. Blumberg, "Development Evolving: The Origins and Meanings of Instinct," National Library of Medicine, https://www.ncbi.nlm.nih.gov/pmc/articles/PMC5182125/.

370 Bryan Johnson, "Top 10 Human Reflexes and Natural Instincts," January 28,2012, https://listverse.com/2012/01/28/top-10-human-reflexes-and-natural-instincts/.

371 J. Joost, L.M. Bierens, Philippe Lunetta, Mike Tipton, and David S. Warner, "Physiology of Drowning: A Review," American Physiological Society, https://journals.physiology.org/doi/full/10.1152/physiol.00002.2015.

372 Melissa Mayntz, "Bird Migration — How It Works," December 24, 2020, https://www.thespruce.com/how-birds-migrate-386445.

373 Priyanka Runwal, "Bird Migration Is One of Nature's Great Wonders. Here's How They Do It," National Geographic, May 5, 2021, https://www.nationalgeographic.com/animals/article/bird-migration-one-of-natures-wonders-heres-how-they-do-it.

374 Melissa Mayntz, "Bird Migration — How It Works," The Spruce, December 24, 2020, https://www.thespruce.com/how-birds-migrate-386445.

375 Melissa Mayntz, "Why Do Birds Migrate?" The Spruce, October 2, 2019, https://www.thespruce.com/why-do-birds-migrate-386453.

376 Ibid.

377 Melissa Mayntz, "12 Types of Bird Migration," The Spruce, September 7, 2019, https://www.thespruce.com/types-of-bird-migration-386055.

378 Priyanka Runwal, "Bird Migration Is One of Nature's Great Wonders ," National Geographic, May 5, 2021, https://www.nationalgeographic.com/animals/article/bird-migration-one-of-natures-wonders-heres-how-they-do-it.

379 Melissa Mayntz, "Bird Migration — How It Works," The Spruce, December 24, 2020, https://www.thespruce.com/how-birds-migrate-386445.

380 Priyanka Runwal, "Bird Migration Is One of Nature's Great Wonders," National Geographic, May 5, 2021, https://www.nationalgeographic.com/animals/article/bird-migration-one-of-natures-wonders-heres-how-they-do-it.

381 Melissa Mayntz, "Bird Migration — How It Works," The Spruce, December 24, 2020, https://www.thespruce.com/how-birds-migrate-386445.

382 Ibid.

383 Priyanka Runwal, "Bird Migration Is One of Nature's Great Wonders," National Geographic, May 5, 2021, https://www.nationalgeographic.com/animals/article/bird-migration-one-of-natures-wonders-heres-how-they-do-it.

384 Ibid.

385 "Choosing the route," RSPB, https://www.rspb.org.uk/birds-and-wildlife/natures-home-magazine/birds-and-wildlife-articles/migration/choosing-the-route/.

386 Ibid.

387 Ibid.

388 Priyanka Runwal, "Bird Migration Is One of Nature's Great Wonders," National Geographic, May 5, 2021, https://www.nationalgeographic.com/animals/article/bird-migration-one-of-natures-wonders-heres-how-they-do-it.

389 "On the Move," https://www.rspb.org.uk/birds-and-wildlife/natures-home-magazine/birds-and-wildlife-articles/migration/on-the-move/.

390 Priyanka Runwal, "Bird Migration Is One of Nature's Great Wonders," National Geographic, May 5, 2021, https://www.nationalgeographic.com/animals/article/bird-migration-one-of-natures-wonders-heres-how-they-do-it.

391 "On the Move," https://www.rspb.org.uk/birds-and-wildlife/natures-home-magazine/birds-and-wildlife-articles/migration/on-the-move/.

392 Melissa Mayntz, "Bird Migration — How It Works," The Spruce, December 24, 2020, https://www.thespruce.com/how-birds-migrate-386445.

393 Ibid.

394 "How Do Birds Navigate?" Resource Library, https://www.nationalgeographic.org/media/how-do-birds-navigate/.

395 "Bird Migration," https://www.rspb.org.uk/birds-and-wildlife/natures-home-magazine/birds-and-wildlife-articles/migration/.

396 Melissa Mayntz, "Bird Migration — How It Works"December 24, 2020, https://www.thespruce.com/how-birds-migrate-386445.

397 "On the Move," RSPB, https://www.rspb.org.uk/birds-and-wildlife/natures-home-magazine/birds-and-wildlife-articles/migration/on-the-move/.

398 "Choosing the Route ," RSPB, https://www.rspb.org.uk/birds-and-wildlife/natures-home-magazine/birds-and-wildlife-articles/migration/choosing-the-route/.

399 Melissa Mayntz, "Bird Migration — How It Works," The Spruce, December 24, 2020, https://www.thespruce.com/how-birds-migrate-386445.

400 Priyanka Runwal, "Bird Migration Is One of Nature's Great Wonders. Here's How They Do It ," National Geographic, May 5, 2021, https://www.nationalgeographic.com/animals/article/bird-migration-one-of-natures-wonders-heres-how-they-do-it.

401 "The Types and Stages of Insect Metamorphosis" Debbie Hadley, July 26, 2019, https://www.thoughtco.com/types-of-insect-metamorphosis-1968347.

402 "Butterfly Life Cycle," Florida Museum, August 22, 2022, https://www.floridamuseum.ufl.edu/educators/resource/butterfly-life-cycle.

403 "Metamorphosis," BD Editors, October 4, 2019, https://biologydictionary.net/metamorphosis/

404 Ibid.

405 Ibid.

406 Ibid.

407 Ibid.

408 Ibid.

409 Ibid.

410 Ibid.

411 "The Life Cycle of a Salmon," Animal Metamorphosis, May 14, 2012, http://animalia-metamorphosis.blogspot.com/2012/05/life-cycle-of-salmon.html.

412 "How Do Salmon Know Where Their Home Is When They Return from the Ocean?" USGS, https://www.usgs.gov/faqs/how-do-salmon-know-where-their-home-when-they-return-ocean.

413 "Spawning Salmon Don't Eat," Oregon Fishing Forum, September 12, 2011, https://www.oregonfishingforum.com/threads/spawning-salmon-dont-eat.20624/.

414 Flora Graham, "Daily Briefing: Sea Slugs Cut Off Own Heads to Grow a Fresh Body," Nature, March 10, 2021, www.nature.com/articles/d41586-021-00631-w.

415 Ibid.

416 "Stretching — an Overlooked Instinct," https://sharingthehealth.com/2017/03/15/stretching-an-overlooked-instinct/.

417 Dr. Michael Denton, Evolution: A Theory in Crisis (Chevy Chase, MD: Adler & Adler, 1986), p. 202–203.

418 John Gibbons, "Keeping Warm in Winter Is for the Birds," Smithsonian, January 30, 2015, https://www.si.edu/stories/keeping-warm-winter-birds.

419 "Dormancy," Sources Select Resources, https://www.sources.com/SSR/Docs/SSRW-Dormancy.htm.

420 "Ask a Naturalist: Hibernation vs. Brumation vs. Estivation," Discovery Place, January 13, 2016, https://nature.discoveryplace.org/blog/ask-a-naturalist-hibernation-vs.-brumation-vs.-estivation.

421 "Animal Instincts: 9 of the World's Best Wildlife Experiences," Wanderlust, September 9, 2018, https://www.wanderlust.co.uk/content/breathtaking-wildlife-encounters/.

422 "Instinct: Definition & Explanation," https://study.com/academy/lesson/instinct-definition-lesson-quiz.html.

423 Leslie P. Kozak, Martin E. Young, "Heat from Calcium Cycling Melts Fat," Nature Medicine, https://www.nature.com/articles/nm.2956.

424 Christie Wilcox, "Some Animals Don't Actually Sleep for the Winter, and Other Surprises About Hibernation," National Geographic, https://www.nationalgeographic.com/animals/article/animals-hibernation-science-nature-biology-sleep.

425 "Hibernating Mammals and Brumating Reptiles: What's the Difference?" January 20, 2014, http://infinitespider.com/hibernating-mammals-brumating-reptiles-whats-difference/.

426 Sophie Weiner, "How Do Non-human Animals Experience Hunger?" May 29, 2017, https://gizmodo.com/how-do-animals-experience-hunger-1795337295.

427 "What Is Estivation? Animals That Estivate," https://www.worldatlas.com/articles/what-is-estivation-animals-that-estivate.html.

428 Colin Beer, "Instinct," Britannica, https://www.britannica.com/topic/instinct.

429 Marc Bekoff, "Animal Instincts: Not What You Think They Are," Greater Good Magazine, March 8, 2011, https://greatergood.berkeley.edu/article/item/animal_instincts.

430 Kenneth Maiese, MD, "Overview of the Nervous System," https://www.merckmanuals.com/home/brain,-spinal-cord,-and-nerve-disorders/biology-of-the-nervous-system/overview-of-the-nervous-system.

431 Dominique Debanne, et al., "Axon Physiology," https://hal-amu.archives-ouvertes.fr/hal-01766861/file/Debanne-Physiol-Rev-2011.pdf.

432 E. Luders, P.M. Thompson, A.W. Toga, The Journal of Neuroscience, "The Development of the Corpus Callosum in the Healthy Human Brain," https://www.ncbi.nlm.nih.gov/pmc/articles/PMC3197828).

433 Karlsruher Institut für Technologie (KIT), "Navigation System of Brain Cells Decoded," https://www.sciencedaily.com/releases/2017/10/171025105041.htm

434 Mun Fei Yam, et al., "General Pathways of Pain Sensation and the Major Neurotransmitters Involved in Pain Regulation," https://www.ncbi.nlm.nih.gov/pmc/articles/PMC6121522/.

INDEX: